Biomass Energy Economics and Rural Livelihood in Sichuan, China

DEVELOPMENT ECONOMICS AND POLICY

Series edited by Franz Heidhues †, Joachim von Braun,
Ulrike Grote and Manfred Zeller

Vol. 78

Biomass Energy Economics and Rural Livelihood in Sichuan, China

Qiu Chen

Bibliographic Information published by the Deutsche Nationalbibliothek
The Deutsche Nationalbibliothek lists this publication in the Deutsche Nationalbibliografie; detailed bibliographic data is available in the internet at http://dnb.d-nb.de.

Library of Congress Cataloging-in-Publication Data
Names: Chen, Qiu, author.
Title: Biomass energy economics and rural livelihood in Sichuan, China / Qiu Chen.
Description: New York : Peter Lang, [2017] | Series: Development economics and policy, ISSN 0948-1338 ; Vol. 78 | Includes bibliographical references.
Identifiers: LCCN 2017044250| ISBN 9783631739235 (Print : alk. paper) | ISBN 9783631739525 (E-PDF) | ISBN 9783631739532 (EPUB) | ISBN 9783631739549 (MOBI)
Subjects: LCSH: Biomass energy--China--Sichuan Sheng. | Households—Energy consumption--China--Sichuan Sheng.
Classification: LCC HD9502.5.B543 C45 2017 | DDC 333.95/39095138--dc23 LC record available at https://lccn.loc.gov/2017044250

ISSN 0948-1338
ISBN 978-3-631-73923-5 (Print)
E-ISBN 978-3-631-73952-5 (E-PDF)
E-ISBN 978-3-631-73953-2 (E-Pub)
E-ISBN 978-3-631-73954-9 (mobi)
DOI 10.3726/b12724

© Peter Lang GmbH
Internationaler Verlag der Wissenschaften
Berlin 2018
All rights reserved.
Peter Lang – Berlin · Bern · Bruxelles · New York ·
Oxford · Warszawa · Wien

All parts of this publication are protected by copyright. Any utilisation outside the strict limits of the copyright law, without the permission of the publisher, is forbidden and liable to prosecution. This applies in particular to reproductions, translations, microfilming, and storage and processing in electronic retrieval systems.

This publication has been peer reviewed.

www.peterlang.com

Abstract

This study investigates the influences of household biomass energy use on rural livelihoods in Sichuan Province of China. Of the 556 surveyed households, 432 (77.7%) households still use traditional solid biomass energy (crops straw and firewood) for cooking, while 243 (43.7%) households produce biogas. An alternative-specific conditional logit model was adopted to test the determinants of household biomass energy choice behaviors from the perspectives of households' revealed preferences and stated preferences (RP and SP) based on the random utility theory. The results of this study indicated that households prefer to use energy sources with lower prices (costs), higher safety, and lower indoor pollution. Moreover, this study showed that the decision maker characteristics, the demographic structure of rural families, income level, arable land owned and household location are all crucial factors affecting the process of household energy transition.

On the production side, in order to investigate the influence of traditional biomass energy use on agricultural production, a multioutput profit function was adopted to further analyze the relationship between agricultural production and biomass collection. The estimation results showed that the supply cross-price elasticities of agricultural products and biomass are −0.02 and −0.52, respectively, indicating that biomass collection could bring a negative effect to agricultural production due to the competition between these two activities for limited labor resources.

Finally, this research provided a holistic and comprehensive analysis of household biomass energy using behaviors based on an agricultural household model. The estimation results revealed that household biomass energy consumption responds positively to the changes in exogenous prices of self-consumed agricultural products and labor, while the market failures reduce the flexibility of household biomass energy using behaviors in the cases of changes in the price of commercial energy or other marketed goods.

Zusamenfassung

Diese Studie untersucht die Auswirkungen der Nutzung von Biomasse Energieverbrauch auf ländliche Lebensgrundlagen in der Provinz Sichuan in China. Von den 556 befragten Haushalte, 432 (77,7%) Haushalte verwenden noch traditionelle feste Biomasse (Stroh und Holz) zum Kochen, 243 (43,7%) Haushalte wählen Biogas zu erzeugen. Ein alternativ-spezifisches bedingtes Logit Modell wird angewendet auf der Grundlage der Zufallsnutzentheorie, um die Determinanten des Wahlverhaltens der Biomasse aus den Perspektiven der offenbarten Präferenzen und der geäußerter Präferenzen zu testen. Die Ergebnisse dieser Studie zeigen, dass die Haushalte gern die Energie mit niedrigeren Preisen (kosten), höhere Sicherheit und geringere Innenraumverschmutzung verwenden. Darüber hinaus zeigt diese Studie, dass die Eigenschaften der Entscheidungsträger; die demografische Struktur; Einkommensniveau; der Besitz der Ackerland und der Haushaltsstandort sind alle entscheidenden Faktoren, die den Prozess der Transformation der Hausenergie beeinflussen.

Um die Auswirkungen der traditionellen Nutzung von Biomasse auf die landwirtschaftliche Produktion zu untersuchen wurde mit Multi-Output Gewinn Funktion die Beziehung zwischen der landwirtschaftlichen Produktion und Biomasse weiter analysieren. Die Ergebnisse zeigen, dass die Kreuzpreiselastizitäten des Angebots zwischen Agrarerzeugnisse und Biomasse wurden −0,02 und −0,52. Es zeigte sich, dass die Biomasse Sammlung einen negativen Einfluss auf die landwirtschaftliche Produktion haben können. Das führt einen Wettbewerb zwischen diesen beiden Tätigkeiten für begrenzte Arbeitsressourcen.

Schließlich stellt diese Studie eine ganzheitliche und umfassende Analyse der Nutzung von Biomasse zur Energieerzeugung mit einem landwirtschaftlichen Haushaltsmodell. Die Ergebnisse zeigen, dass der Energieverbrauch der Haushalte aus Biomasse eine positive Reaktion auf die Veränderungen der exogenen Preise von selbstverbrauchenden Agrarprodukten und Arbeitskräften hat. Und in den Fällen von Änderungen des Preises der kommerziellen Energie oder anderer vermarkteter Güter Marktversagen reduziert die Flexibilität der Haushalts Biomasse-Energie.

Table of Contents

Abstract 5

Zusamenfassung 7

List of Tables 13

List of Figures 15

Acknowledgements 17

Abbreviations 19

Chapter 1   Introduction 21

1.1   Research Background 21

1.2   Problem Statement 22

1.3   Context of the Study Region 24

    1.3.1   Socioeconomic Status of Rural Sichuan 24

    1.3.2   Status of energy consumption in rural Sichuan Province 28

    1.3.3   Policy Background for Rural Biomass Energy
Construction in Sichuan 30

1.4   Research Objectives and Questions 32

1.5   Conceptual Framework 33

1.6   Data 34

    1.6.1   Sampling procedures 34

    1.6.2   Sample description..35

1.7   Main Contributions of the Thesis 38

1.8   Organization of the Thesis 38

Chapter 2   Analytical Framework 41

2.1   An Agricultural Household Model 41

2.2 Household Energy Choice and Its Determinants 43
2.3 The Impacts of Biomass Collection on Agricultural Production 47
2.4 The Impacts of Household Biomass Energy Utilization Behaviors on Rural Livelihood 50
 2.4.1 Impacts of Biomass Energy Utilization on Rural Livelihood in a Separable Agricultural Household Model 51
 2.4.2 Impacts of Biomass Energy Utilization on Rural Livelihood in a Non-separable Agricultural Household Model 53
2.5 Conclusion 58

Chapter 3 Household Biomass Energy Choice for Cooking in Energy Transition and Its Impacts on Rural Livelihoods 61

3.1 Introduction 61
3.2 Literature Review on Economics of Energy Choices 62
3.3 Household Cooking Energy Use in Study Region 64
3.4 Empirical Strategy 71
 3.4.1 Revealed and Stated Preferences 71
 3.4.2 Model Specification 74
3.5 Empirical Analysis 75
 3.5.1 Revealed Preference (RP) Analysis 75
 3.5.2 Stated Preference (SP) Analysis 85
 3.5.3 Joint estimation of RP-SP data 94
3.6 Conclusion 97

Chapter 4 Evaluating the Impacts of Biomass Collection on Agricultural Production 99

4.1 Introduction 99
4.2 Literature Review 100
4.3 Biomass Collection and Its Impacts on Agricultural Production in Study Region 101
 4.3.1 Research Context in Sichuan Province 101

  4.3.2 Biomass Collection and Agricultural Production: Preliminary Data Analysis 102

4.4 Empirical Analysis 108
  4.4.1 Estimating Shadow Wage and Shadow Price 108
  4.4.2 Estimating Profit Function 119

4.5 Conclusion 124

Chapter 5 Impacts of the Changes in Exogenous Markets on Household Biomass Energy Use 127

5.1 Introduction 127

5.2 Descriptive analysis 128
  5.2.1 Context of Household Energy Consumption 128
  5.2.2 Varable Description 131

5.3 Separability 133
  5.3.1 Model Specification 133
  5.3.2 Data and Estimation 135

5.4 Household Behavior Analysis 138
  5.4.1 Shadow Wage and Shadow Price Estimation 139
  5.4.2 Household Consumption Decisions 143
  5.4.3 Household Labor Supply 151

5.5 Household Biomass Energy Use Responses to the Changes in Exogenous Markets 155

5.6 Conclusion 157

Chapter 6 Conclusions and Policy Implications 161

6.1 Conclusions 161
  6.1.1 Biomass Energy Choice in Energy Transition and Its Determinants 162
  6.1.2 Impacts of Biomass Collection on Agricultural Production 162
  6.1.3 Biomass Energy Use Responses of Households to the Exogenous Market 163

6.2 Policy Implications 164

  6.2.1 Adjust Energy Price and Improve Energy Quality 164
  6.2.2 Enhancing Households' Access to Modern Fuels 165
  6.2.3 Eliminating the Market Faliures 165
  6.2.4 Increasing Attention to Regional Differences 166

References 167

Appendix 183

List of Tables

Table 1.1	Selected socioeconomic indicators for Sichuan Province	26
Table 1.2	General characteristics of sampled households	36
Table 3.1	Comparison of characteristics of households choosing four different energy alternatives	69
Table 3.2	Description of explanatory variables used in asclogit model for RP data	79
Table 3.3	Estimation results of asclogit model for RP data	82
Table 3.4	Marginal effects of key influencing factors in RP asclogit model	83
Table 3.5	Assignment of levels and labels for attributes in cooking energy choice experiments	88
Table 3.6	A sample of a choice set in the choice experiment	89
Table 3.7	Description of explanatory variables used in asclogit model for SP data	90
Table 3.8	Estimation results of asclogit model for SP data	93
Table 3.9	Marginal effects of key influencing factors in SP asclogit model	94
Table 3.10	Joint estimation results of the combination of RP and SP data	96
Table 4.1	General information of household participation in agricultural production and biomass collection in study region	103
Table 4.2	General information of households collecting biomass	105
Table 4.3	Descriptive information of household members who are mainly responsible for biomass collection	106
Table 4.4	Description of household characteristics and variables used in model estimation	113
Table 4.5	Multivariate Probit Estimates of household participation functions of agricultural production, biomass collection and off-farm work	116
Table 4.6	Estimation results of the system of production functions using IT3SLS	118

Table 4.7	Data and variables used in estimating SNQ profit function	122
Table 4.8	Estimated price elasticities of outputs and inputs	123
Table 5.1	Energy consumption status of sampled households (Kgsce per year per household)	129
Table 5.2	Traditional biomass energy cousumption of sampled households (Kg per year per household)	129
Table 5.3	Household use of biogas (Total sample size: 524 households)	130
Table 5.4	Commercial energy consumption of sampled households	131
Table 5.5	Socioeconomic characteristics of sampled households	132
Table 5.6	Household time allocation to different activities (hours per year)	133
Table 5.7	Descriptive information about variables in the FMM model	136
Table 5.8	On-farm Labor Allocation: OLS and FMM estimates	138
Table 5.9	Descriptive analysis of data in production system estimation	140
Table 5.10	Estimates of Seemingly Unrelated Probit (SUP) model for biomass collection and market participation for sampled households	141
Table 5.11	Estimation results of the simultaneous equations using IT3SLS	142
Table 5.12	Description of variables used in estimating AL/AIDS model	147
Table 5.13	Parameter estimation of LA/AIDS model using censored SURE	148
Table 5.14	Price and income elasticity estimated by LA/AIDS model (mean value)	150
Table 5.15	Constrained IT3SLS estimation results of labor share equations	154
Table A.1	Estimation results of the normalized quadratic profit function with imposition of convexity	187

List of Figures

Figure 1.1	Location of Sichuan Province	25
Figure 1.2	Rural energy consumption in Sichuan Province (2006–2013)	28
Figure 1.3	Rural residential energy consumption in Sichuan Province	30
Figure 1.4	Rural energy consumption for productive activities in Sichuan Province	30
Figure 1.5	Conceptual framework	33
Figure 1.6	Geographical distribution of study region	35
Figure 1.7	Surveyed households' energy use for residence in study regions	37
Figure 2.1	Pathway of the equilibrium on the production-possibility curve of the household with decrease (Left) and increase (Right) in wage rate	52
Figure 3.1	Cooking fuel ladder in Sichuan Province	64
Figure 3.2	Cooking energy use pattern of surveyed households	65
Figure 3.3	Energy use for cooking of surveyed households	66
Figure 3.4	Income levels of two household groups	67
Figure 3.5	Cooking stoves four main types of energy sources in study region	68
Figure 3.6	Energy choices made by household heads with different gender	69
Figure 3.7	Energy choices made by household heads of different age groups	70
Figure 3.8	Energy choices made by household heads with different educational levels	71
Figure 3.9	Household energy choices (Unit: households)	77
Figure 4.1	Biomass collected in three different regions	102
Figure 4.2	The relationship between biomass collection and agricultural production	107
Figure A.1	Optimal labor allocation in separable AHM	183

Figure A.2 The changes in equilibrium with decrease in wage rate in separable AHM 184

Figure A.3 The changes in equilibrium with increase in wage rate in separable AHM 185

Figure A.4 Optimal labor allocation in non-separable AHM 186

Acknowledgements

Foremost, I would like to express my deepest and sincerest gratitude to my first supervisor, Prof. Dr. Joachim von Braun for his continuous support of my Ph.D study and relaterd research, for his patience, enthusiasm, and immense knowledge. His guidance helped me in all the time of research and writing of this thesis. I would not have completed this theis without his effort.

I would like to thank Dr. Alisher Mirzabaev for his insightful comments and encouragement. He always provides answers and solutions to my questions and problems with his intelligence and knowledge.

My sincere thanks also go to Biogas Institutes of Ministry of Agriculture (BIOMA), Rural Energy Office of Sichuan government (SCREO) and Sichuan Agricultural University (SAU) for their assistance in acquiring first hand data. Speical thanks are due to Prof. Dr. Yuansheng Jiang, Prof. Qichun Hu, Prof. Dr. Wenguo Wang, Dr. Haoran Yang, Dr. Lili Jia, Dr. Tianbiao Liu and Mr. Wanlin He for their enthusiastic support and help.

I would like to thank Dr. Manske and Ms. Retat-Amin for administrative support and help in many aspencts of my life in Bonn. I thank all my colleagues at ZEF for the meaningful discussions and for all the fun we have had in the last four years.

Particularly, I am grateful to China Scholarship Council (CSC) and Dr. Hermann Eiselen Doctoral Program of the Foundation of Fiat Panis for funding my research at Germany.

Last but not the least, I would like to thank my parents and my boyfriend, Liangying Xie, for supporting me spiritually throughout writing this thesis and my life in general.

Abbreviations

AHM	Agricultural household model
ARA	Available resources amount
ASCLOGIT	Alternative-specific conditional logit
BIOMA	Biogas Institute of Ministry of Agriculture
CNY	Chinese Yuan
CNBS	China National Bureau of Statistics
CO_2	Carbon dioxide
CRES	Chinese Renewable Energy Society
DCE	Discrete choice experiment
DFID	Department for International Development
ESPA	Ecosystems Services for Poverty Alleviation
FMM	Finite Mixture Model
GDP	Gross domestic product
GHG	Greenhouse gas
GL	Generalized Leontief
ha	Hectares
IIED	International Institute for Environment and Development
IMR	Inverse Mills Ratio
IT3SLS	Iterative Three-stage Least Squared
Kg	Kilogram
Kgsce	Kilogram of standard coal equivalent
Km	Kilometer
LA/AIDS	Linear Approximation of the Almost Ideal Demand System
LPG	Liquid petroleum gas
NEA	National Environment Agency
NQ	Normalized quadratic
OLS	Ordinary Least Squared
RP	Revealed preference
RPT	Rate of product transformation
SCREO	Sichuan Rural Energy Office
SML	Simulated maximum likelihood
SNQ	Symmetric normalized quadratic

SP	Stated preference
SURE	Seemingly unrelated estimation
TL	Translog
TRA	Theoretical resources amount
Tsce	Ton of standard coal equivalent
UNCC	United Nations Climate Change
WTP	Willingness to pay

Chapter 1 Introduction

1.1 Research Background

To date, biomass energy is still one of the most important energy sources used in developing countries, constituting 35% of their energy supply (Demirbas and Demirbas, 2007). It appears to be an attractive type of energy for its renewable, positive environmental properties and its significant economic potential compared to the fossil fuels, which face increasing prices in the future (Cadenas and Cabezudo, 1998). As a kind of renewable resource derived from biological materials, biomass energy links the natural environment and humans' activities. Hence, it is one of the essential elements of the rural livelihood framework. In the case of China, biomass energy is the principal type of energy utilized for livelihood purposes such as cooking, space heating (cooling), and lighting (Wang and Qiu, 2009), occupying approximately 41.8% in the total rural residential energy consumption (CRES, 2011). As a large agricultural country of the world, China has abundant biomass resources, including agricultural residues, woody biomass, animal dung, and urban living waste from which to produce multiple forms of energy in substantial quantities across a wide distribution area (Zhang et al., 2010). Nearly 89.4% of these resources are consumed as fuel for residential purposes (CRES, 2011). Despite the important role that biomass energy plays in the agriculture-based rural livelihood system, due to the limited access to advanced energy technologies and modern energy services, a considerable share of the rural population still depends heavily on direct combustion of biomass for domestic utilization. Moreover, in recent years, although the program of rural electrification has obtained great progress, many rural households, especially those living in remote areas, still cannot afford the high prices of electricity attributed to the backward economic development. Traditional biomass energy such as crops, straw, and firewood occupies the dominant place in rural energy consumption. The traditional use of biomass energy has brought many serious consequences to rural lives; for instance, resource waste, indoor air pollution, rural environmental deterioration, and social inequality (Zhang et al., 2010). With the fast rural economic growth, these problems are getting worse. In stark contrast, the widespread use of clean, low-cost, and high-efficiency biomass energy based on modern technologies could significantly improve living standards by providing environmental benefits and generating employment in rural areas (Zhang et al., 2009). Therefore, it is vitally necessary for China to optimize the contributions of biomass energy to rural sustainable development.

With the rise in awareness of the importance of biomass energy to the general rural livelihood, the Chinese central government has emphasized biomass energy construction as an essential component of its long-term rural development strategy. The National Energy Administration (NEA) has also set the promotion of modern biofuels as a core task of the twelfth Five-Year Plan (2011–2015) for Chinese national energy development, with a target of utilizing more than 50 million tsce[1] biomass energy annually by 2015. Nevertheless, coal and some other types of fossil energy such as LPG and natural gas still take up a large proportion in current energy consumption structure resulting in serious threats on Chinese sustainable development. It is rather urgent for China to face the 'energy crisis' with the increasing demand for energy and the national situation of energy supply shortage caused by the depletion of fossil resources. During the United Nations Climate Conference (UNCC) held in December 2009, the Chinese government announced that it will reduce the carbon dioxide emitted per unit of economic output by 40%–45% by 2020 compared to its 2005 levels (Zhang et al., 2009). Biomass energy is considered to be the best substitute for fossil energy and thus is a good measure to quell the sharp conflict between economic growth and environmental protection by the Chinese government. It is also an inevitable strategic option for promoting sustainable rural development. Under this circumstance, study of the impacts of biomass energy utilization on rural livelihood in China has great significance.

1.2 Problem Statement

In recent years, a large number of studies have tried to clarify the impacts of household energy use patterns on rural livelihoods (Gupta, 2003; Cherni et al., 2007; Byrne et al., 2007; Cherni and Hill, 2009; Fan et al., 2011; Lee et al., 2015; Biggs et al., 2015). These studies have shown that energy transition can change rural livelihoods in various ways. Livelihoods comprise the assets, capabilities, and activities required for a means of living (Chambers and Conway, 1992; Bebbington, 1999; DFID, 2000; Scoones, 2009). Not only are economic factors for survival such as income critical for livelihoods, but also non-economic ones that mediate access to different assets such as health status and environmental impacts (Ellis, 2000; Hunsberger et al., 2014). Particularly as a bridge between nature and human-being activities, biomass energy plays a vital role in rural livelihoods. In accordance with the viewpoints of Chamdimba (2009), biomass

---

1   tsce is the abbreviation of the standard energy unit: ton of coal equivalent. Kgsce is kilogram coal equivalent.

energy use is identified as two competing paradigms in rural areas of developing countries. Poor households living in geographically remote areas depend heavily on traditional biomass for most of their basic livelihood activities such as cooking and heating. Conventional burning of biomass with low thermal efficiency usually causes waste of resources (Chen et al., 2006). Additionally, it could also negatively impact human health through indoor air pollution (Fan et al., 2011). Moreover, households who have to spend more time on biomass collection and energy preparation are usually constrained from engaging in income-generating activities (van der Kroon et al., 2013). On the other hand, with fast development of the rural economy, household demands for new efficient and environment-friendly biomass energy are gradually increasing (Chamdimba, 2010). Adopting modern biomass energy could bring positive effects on rural livelihoods by improving quality of life and enhancing environmental protection (Gosen et al., 2013). It is thus pivotal to promote the biomass energy transition. Research on household fuel switching has also indicated that shifts from inferior traditional biomass towards more efficient commercial energy such as electricity could also help increase livelihood outcomes (Hunsberger et al., 2014; Lee et al., 2015). Although the use of modern energy sources at the household level is expected to ensure livelihood sustainability, the understanding of biomass energy use in energy transition and its policy implications for livelihoods are still limited. Furthermore, little attention has been paid to this issue in rural China. Therefore, this paper uses a Chinese case study to investigate household biomass energy utilization and its impact on rural livelihoods in rural areas.

Currently in rural China, traditional use of biomass energy still takes an important position in rural residential energy consumption structure, especially in western areas (Gan and Yu, 2008). As it could bring negative impacts on rural livelihoods, there is no doubt that, along with the increase in the demand of rural households for cleaner energy, the transition shift towards the use of modern biofuels becomes necessary. For this reason, it is essential to inspect the energy choice behaviors of rural households in the process of the transition. However, despite the many empirical studies testing the determinants of household biomass energy choice behaviors in China (Démurger and Fournier, 2011, Fan et al., 2011; Wang et al., 2012; Ping et al., 2012), very few of them scrutinize how these influencing factors affect households' potential preferences for biomass energy, especially for some government-led projects such as biogas. Combining the revealed and stated preferences of households, it is important to examine the determinants of household choice on both actual and hypothetical energy alternatives.

Moreover, it is well known that the biomass collected for household energy use is mainly from agricultural waste. Nevertheless, little concern has been given to the impacts of biomass collection on agricultural production. Actually, due to the fact that biomass collection has a direct and close relationship with household livelihoods within the competition between it and agricultural production for labor resources, the changes in the production system could make this competition become fiercer. Thus, it is worth studying the impacts of biomass collection on agricultural production focusing on labor allocation.

Finally, as the market of biomass energy is nearly absent in rural China, households usually use the biomass collected by themselves. Under this situation, in order to better understand household biomass energy using behaviors, the linkages between biomass collection, agricultural production, biomass consumption, and other markets must be investigated. This also requires a systematic analysis jointly considering both consumption and production decisions on biomass energy utilization. Hence, this research will take advantage of an agricultural household model to analyze the household biomass energy utilization behaviors with concern for livelihood enhancement.

1.3 Context of the Study Region

1.3.1 Socioeconomic Status of Rural Sichuan

Sichuan Province is located in Southwestern China; it borders Tibet Autonomous Region to the west, Qinghai to the northwest, Gansu to the north, Shaanxi to the northeast, Chongqing to the east, Yunnan to the south, and Guizhou to the southeast (See Figure 1.1). It covers an area of approximately 0.485 million km^2 (62.7% of which is ethnic autonomous regions) and consists of two geographically distinct parts within its borders. The eastern part is mostly within the fertile Sichuan basin surrounded by hilly areas, while the western part has numerous mountains that form the easternmost part of the Qinghai-Tibet Plateau. The complicated geographic conditions of Sichuan Province constrain its economic development and make Sichuan one of the key target provinces of the anti-poverty projects of the China Western Development Strategy. 36 of its 183 counties have been officially defined as national poverty counties who qualify for financial support offered by the Chinese central government.

Figure 1.1 Location of Sichuan Province

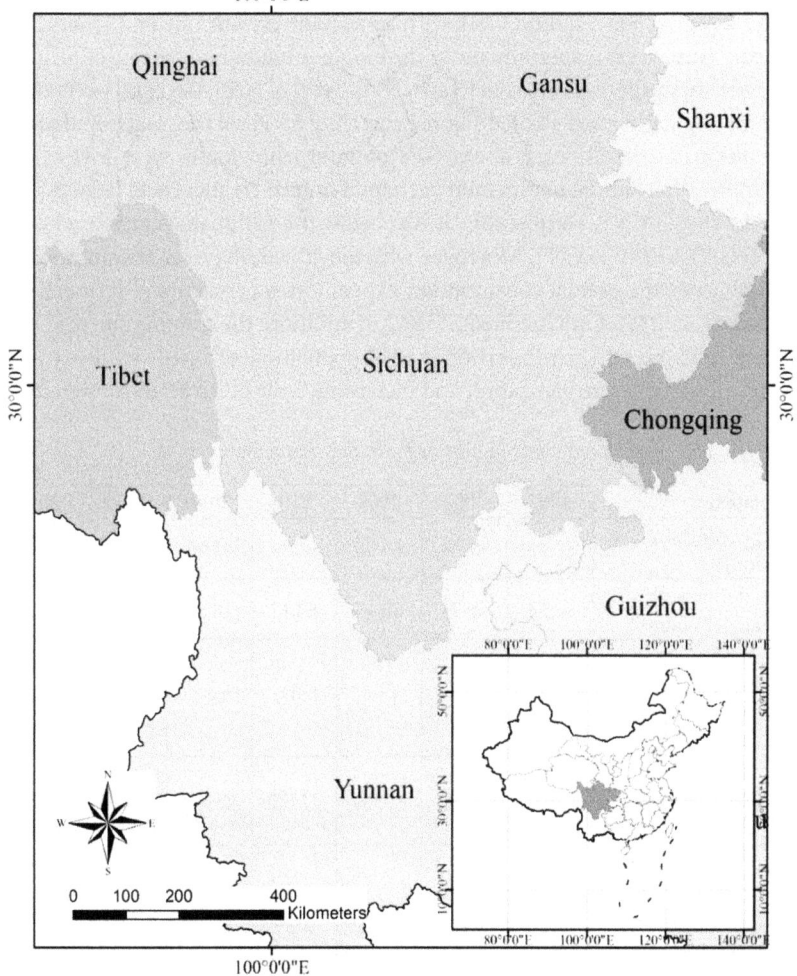

Agriculture is one of the most important sectors in Sichuan Province. According to the historical data from 2007 to 2013 (See Table 1.1), though the share of agriculture (including forestry, animal husbandry and fishery)[2] value added in

---

2 Defined by China National Bureau of Statistics (2003): Regulation on Classification of Three Sectors.

GDP was never more than 20%, at least 40% of the employed labor forces were engaged in agriculture. The annual agriculture value added increased from 203.2 billion CNY[3] to 342.6 billion CNY with an average growth rate of 3.97%. However, the rural areas, especially those in remote mountainous areas, of Sichuan Province are the poorest areas in China. By the end of 2013, the rural permanent resident population was 44.67 million, occupying 55.1% of the total population.[4] Over the past several years, about 60% of rural labor forces were invested in agriculture. The annual net income per capita of farmers increased from 3002.4 CNY to 7895.3 CNY. Despite this, it was below the national average level from 3587.0 CNY to 7916.6 CNY. Moreover, with the growth in prices of commodities in rural areas, the annual consumption expenditures per capita of farmers also went up from 2395 CNY to 5366.7 CNY. In addition, the number of rural permanent residents and its proportion in total population declined simultaneously as a result of the fast urbanization and increasing scale of rural labor migration.

Table 1.1 Selected socioeconomic indicators for Sichuan Province

Indicators	2007	2008	2009	2010	2011	2012	2013
Rural population (million)[a]	66.75	67.04	66.98	66.46	65.95	65.85	65.00
Rural permanent residents (million)[b]	52.34	50.94	50.17	48.12	46.83	45.61	44.67
The proportion of rural permanent residents (% of total population)[a]	64.4	62.6	61.3	59.8	58.2	56.5	55.1
GDP per capita (calculated at current prices in CNY)[a]	12963	15495	17339	21182	26133	29608	32454
Agriculture, Forestry, Animal Husbandry and Fishery, value added (calculated at current prices in billion CNY)[a]	203.2	221.6	224.1	248.3	298.4	329.7	342.6

---

3 The abbreviation of Chinese currency: Chinese Yuan.
4 Statistical Bulletin of National economy and society development in Sichuan (2013).

Indicators	2007	2008	2009	2010	2011	2012	2013
Agriculture, Forestry, Animal Husbandry and Fishery, value added growth (%)[b]	4.4	3	4	4.4	4.5	4.5	3.6
Agriculture, Forestry, Animal Husbandry and Fishery, value added (% of GDP)[a]	19.2	17.6	15.8	14.5	14.1	13.8	13.0
Agriculture, Forestry, Animal Husbandry and Fishery, value added per agricultural worker (calculated at current prices in CNY)[a]	8967.3	10138.2	10450.6	11919.8	14603.5	16560.5	17515.2
Employment in primary industry (% of total employed labor forces)[a]	47.9	46.1	45.1	43.7	42.7	41.5	40.6
Rural employment in primary industry (% of total rural labor forces)[a]	65.6	63.4	62.5	61.0	60.3	59.1	58.3
Annual per capita net income of rural households (CNY)[c]	3546.7	4121.2	4462.1	5086.9	6128.6	7001.4	7895.3
Annual per capita expenditure for consumption of rural households (CNY)[c]	2747.3	3127.9	4141.4	3897.5	4675.5	5366.7	6126.8
Arable land (Mu per person)[c]	1.03	1.03	1.02	1.08	1.15	1.14	0.97
Human Development Index value[d]	-	0.763	-	0.662	-	-	-
Human Development Rank in 31 provinces of China[d]	-	27	-	23	-	-	-

Data sources: a. Sichuan Statistical Yearbook (2007–2013) [in Chinese]; b. Statistical Bulletin of National Economic and Social Development in Sichuan Province (2006–2013)[in Chinese]; c. China Statistical Database, http://data.stats.gov.cn/ [in Chinese]; d. UNDP, China Human Development Report [in Chinese] (2009/10, 2013)

1.3.2 Status of energy consumption in rural Sichuan Province

Figure 1.2 demonstrates the changes in the energy consumption structure of the rural Sichuan Province during the period from 2006 to 2013. It can be seen that the total amount of energy consumption in rural areas remains relatively stable, at about 60 million tsce with an annual mean of 59.2 million tsce. Due to a decrease in the number of rural permanent residents, the energy consumption per capita[5] gradually increased from 1137.4 Kgsce in 2006 to 1349.2 Kgsce in 2013 with an average growth rate of 18.6%.

Figure 1.2 Rural energy consumption in Sichuan Province (2006–2013)

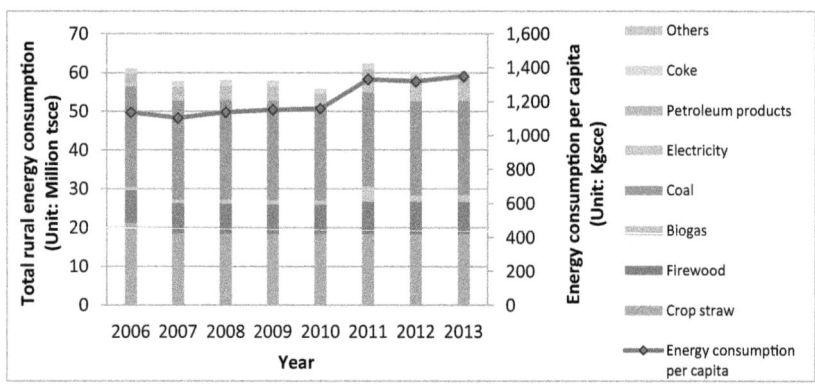

Data sources: Compilation of Sichuan Rural Renewable Energy Statistics (2006–2013)

According to the statistics, coal, crop straw, and firewood are the three main types of energy consumed in rural Sichuan (See Fig. 1.2), accounting for more than 80% of the total. Concretely, coal represents the largest share (approximately 40%) in rural energy consumption. The average consumption amount of coal is about 24.99 million tsce per year. Nevertheless, there is a decline in coal consumption from 25.97 million tsce to 24.29 million tsce during the period between 2006 and 2013. The use of crop straw experienced a decrease during the same period. The yearly mean of crop straw consumption is 18.75 million tsce, occupying around 30% of the total energy consumption in rural areas. However, it does not appear to have obviously affected firewood consumption. The annual

---

5   Calculated by dividing total energy consumption amount in Kgce by rural permanent population.

average firewood consumption is about 8.03 million tsce, accounting for about 14% of the total energy consumption.

In particular, biogas holds a rather small share of rural energy consumption, though the central and local governments have implemented many policies and measures to develop household-based biogas digesters and centralized large-scale biogas system in rural Sichuan. In addition, the consumption of electricity increased from 1.17 million tsce in 2006 to 3.81 million tsce in 2013. Accordingly, its proportion of total energy consumption also grew from 1.9% in 2006 to 6.3% in 2013.

More specifically, energy consumption can be divided into two parts based on use purposes. One is for residence (See Figure 1.3), and the other is for production (See Figure 1.4). Residential energy consumption takes a predominant place, accounting for more than 60% of total rural energy consumption. It can be seen from Figure 1.3 that biomass energy takes the largest share in rural residential energy consumption. Of the total residential energy consumption, 73.2% was biomass energy in 2006. By the end of 2013, it fell to 68.9%. The three main types of biomass energy used by rural households for living purposes are crop straw, firewood, and biogas, among which, crop straw is the most commonly consumed energy source for basic living activities, especially for cooking. Next is firewood, followed by biogas. According to the statistics, the residential consumption of crop straw goes down from 21.23 million tsce in 2006 to 17.07 million tsce in 2013. Nevertheless, it still takes the largest share in the total. The amount of firewood consumed remained relatively steady during the same period, accounting for about 18%-20% of the total residential energy consumption. The biogas consumption, which has been vigorously promoted by both central and local governments, fluctuated during these 8 years and reached its peak in 2011.

On the other side, the production in rural Sichuan still depends heavily on coal. In the production sector, the proportion of biomass energy is quite small, accounting for no more than 5% of the total consumption amount before 2010. In 2010, crop straw began to be used for productive activities with the development of gasification and briquetting technologies in pilot. At that point, the proportion of biomass energy consumption doubled but was still insignificant.

Figure 1.3 Rural residential energy consumption in Sichuan Province

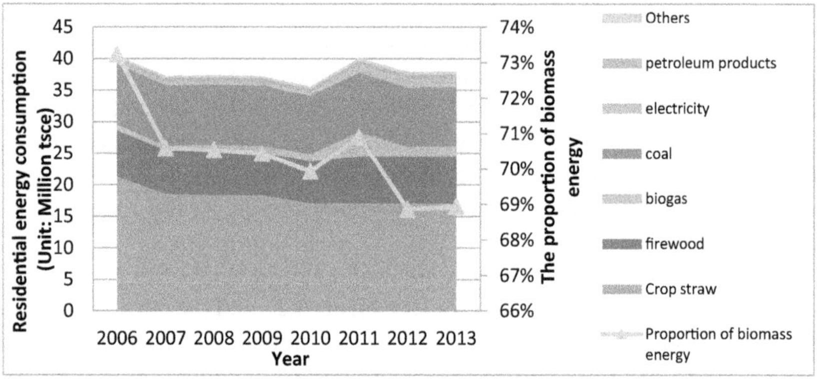

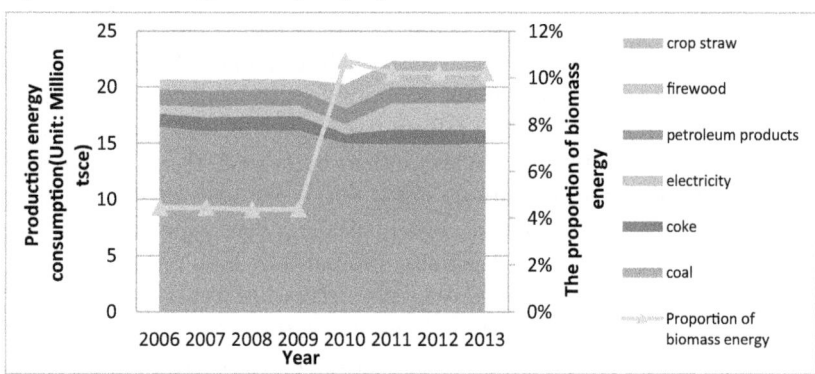

Sources: Compilation of Sichuan Rural Renewable Energy Statistics (2006–2013)

Figure 1.4 Rural energy consumption for productive activities in Sichuan Province

Sources: Compilation of Sichuan Rural Renewable Energy Statistics (2006–2013)

1.3.3 Policy Background for Rural Biomass Energy Construction in Sichuan

As an important agricultural province in China, Sichuan Province has abundant biomass resources derived from agricultural production. Over the past several years, many policies, strategies, and planning documents have been released by central and local governments for the development of biomass energy in Sichuan.

- In 2000, the Ministry of Agriculture (MOA) designed the Prosperous Eco-farmyards plan, which aims to simultaneously improve rural living conditions and reduce the environmental pollution caused by traditional energy use (Chen et al., 2010; Gosens et al., 2013). A nationwide construction boom of household-based biogas digester started with this plan.
- In 2004, the China National Energy Strategy and Policy 2020 (NESP) was published by National Development and Reform Commission (NDRC, 2004). It provided Chinese strategies, objectives, and measures for renewable energy development. For the particular target for decreasing traditional using of biomass, it proposed that, with an annual decrease rate of 2 percent, the traditional use of biomass in rural areas will decrease to 140 million tsce in 2020.
- With regard to energy development, China's Agenda 21 (1997; 2002) emphasized the importance of ensuring national energy supply in ways that safeguard public health and environment in order to achieve an equitable distribution of modern energy services throughout the nation.
- In 2007, the National Rural Biogas Construction Plan (2006–2010), the Development Plan for the Agricultural Bioenergy Industry (2007–2015) (MOA, 2007a; 2007b), and the Medium- and Long-term Development Plan for Renewable Energy in China (NDRC, 2007) were released to promote biomass energy construction in rural areas, especially for biogas construction with the 2010 target of 40 million household-based digesters and 4700 medium- and large-scale biogas plants annually producing 4 billion m^3 biogas as well as the 2020 target of 80 million household biogas digesters and 10000 plants generating 14 billion m^3 biogas per year. Additionally, the capacity for biomass power generation is targeted to reach 30 GW by 2020 (MOA, 2007b).
- The Renewable Energy Law (REL) (Issued in 2005 and Amended in 2009) is the cornerstone of renewable energy development in China aiming at the promotion of the use of renewable energy, the adjustment of energy consumption structure, the protection of environment, and the enhancement of energy security. The newly approved REL also stipulates the establishment of a special fund for developing renewable energy.
- The 12th (2011–2015) Five-Year Plans for National Economic and Social Development established the following initiatives as central to the energy policy in China: Developing clean and safe energy; Renewable energy technologies should be promoted by providing financial support, including taxation, subsidization, and investment policies.

In Sichuan Province, many additional national and local energy policies are listed as follows:

- Since 2011, the standard of subsidy for household biogas digester construction leveled up to 2000 CNY per digester, with the support of the contributions from the national bonds, while the subsidy from the local government should not be less than 500 CNY per digester (NRDC/MOA, 2011).
- Since 2011, the standard of subsidy for building large-scale biogas plants in Sichuan Province is 1500 CNY per m^3 biogas production capacity, whereas the subsidy standard for the small biogas projects is 0.06 million CNY per project from the central government, with more than 0.036 million CNY per project from the local government.
- In 2015, Sichuan has been approved to construct 29 large-scale biogas plants (Yong, 2015) with 123.92 million in CNY subsidies from the central government.

1.4 Research Objectives and Questions

For the purpose of investigating the impacts of biomass energy utilization on rural livelihood, this thesis poses the following main research question:

Does biomass energy utilization influence rural livelihoods in China?

Taking Sichuan Province as an example, this thesis aims to gain better understanding of the linkages between biomass energy utilization and rural livelihoods through answering the following subquestions:

1.1 What core factors impact household cooking fuel choice behaviors?
This study examines the biomass energy choice behaviors of rural households in the process of energy transition and tries to test the determinants of these behaviors and thus determine the way to encourage households to switch from the use of traditional solid biomass energy to clean and effective energy alternatives.

1.2 Does biomass collection influence agricultural production?
This study seeks to identify the impacts of biomass collection on agricultural production.

1.3 Do the changes in exogenous markets (including energy market, labor market, and agricultural products market) affect household biomass energy use?
This study seeks to clarify the household decision-making behaviors of biomass energy utilization and to evaluate the effects of the price changes in exogenous

markets such as energy market, labor market, and agricultural products market on household biomass energy consumption.

1.5 Conceptual Framework

The conceptual framework of this research is presented in Figure 1.5. According to our research question in Section 1.4, the arrows represent direct relations between components. The Q1, Q2, Q3, and Q4 denote the three sub-questions (1.1–1.4) of our research. In our research, households make choices on different energy sources based on some important drivers (determinants). If they decide to use biomass energy, they must decide how much biomass energy they consume to meet their energy demand and how to allocate their limited labor resources to biomass collection and agricultural production. All of these behaviors will impact rural livelihoods. The drivers are selected on the basis of literature review. They mainly consist of household characteristics (household demographic structure, decision maker characteristics, household location, and income level), prices of agricultural products, commercial energy and other marketed goods, and other exogenous variables.

Figure 1.5 Conceptual framework

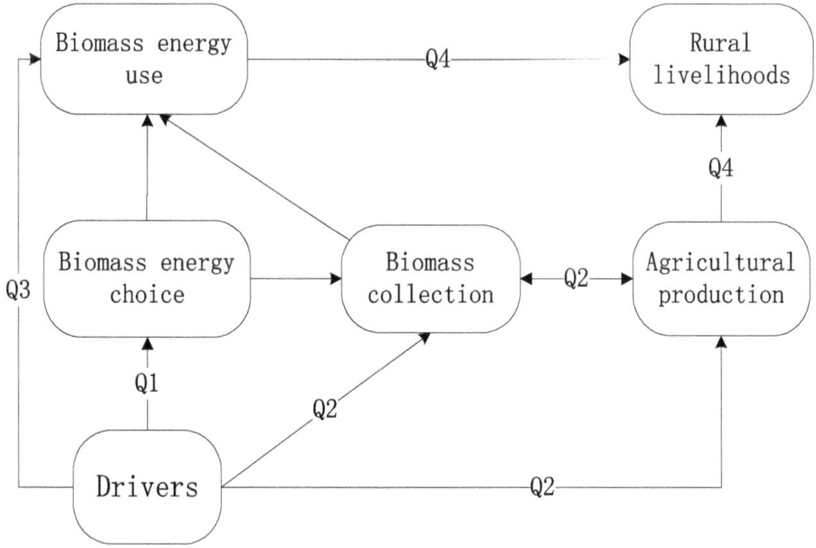

Source: Author's own depiction

1.6 Data

The data used in this research were obtained from a household survey conducted from August 2013 to February 2014 in Sichuan Province. Our sampling methods are described as follows:

1.6.1 Sampling procedures

The sample selection was done in a way that guarantees the representativeness of the overall population. This follows a 95% confidence interval and a margin of error of 5%. To determine our sample size, we applied the following formula to make the calculation:

$$n = \frac{N}{1 + NE^2}$$

Where N denotes the total number of the rural population of Sichuan Province; n is the sample size, and E is the sampling error.

The sample size should be 384 for Sichuan Province, which can be determined using the calculator given in the website http://www.surveysystem.com/sscale.htm, with the formula listed above. To compensate for any missing or failed cases, the sample size is determined to be at least 400 respondents. More specifically, 176 counties of Sichuan Province are sorted by the rural per capita net income level. Then the province is divided into three zones – high, middle, and low – in light of the income levels of rural areas. In each zone, two counties are randomly selected. Furthermore, three towns, each with two villages, are randomly selected in each county. In every village, 15–16 respondents are randomly surveyed. In total, the number of respondents should be 540–576.

Actually, in practice, once the number of households to be surveyed from each village is determined, a plan for selecting households then has to be designed (CDC, 2008). To get the sampling interval, i.e. the space between every two selected households, we divide the total number of households in a village by our subsample size.[6] Thus, for a village of 150 households and a subsample size of 15, every 10th house would be interviewed. In mountainous areas, the sampling interval is too large to create long distances for surveyors to travel between houses. To deal with this problem, we divide our survey areas into 2 sub-areas, and we visit every 5th house to cover half of the sub-areas to reduce the sampling

---

6   In our study region, the average population of one village is about 150. Here, in order to simplify calculation, we assume the subsample size is 15.

interval. In order to reduce the bias of household selection, we randomly select our first visited household. The sampling interval is then adopted to select subsequent households.

1.6.2 Sample description

After eliminating invalid questionnaires and outliers, the total sample size of our study is 556 households. For the low-income zone, two counties, Jiuzhaigou and Mao Counties, are selected from the Aba Prefecture. Counties selected for the middle-income zone are Jiang'an and Changning, located in Yibin City. Finally, Mianzhu and Shifang from Deyang City are selected as representatives for the high-income zone (See Fig. 1.6).

Figure 1.6 Geographical distribution of study region

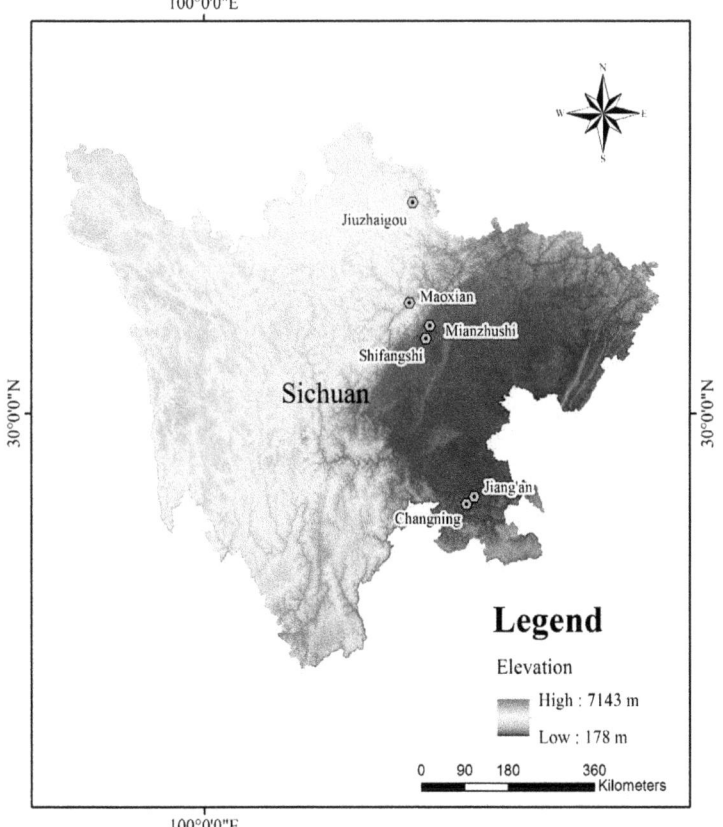

Table 1.2 General characteristics of sampled households

	Aba	Yibin	Deyang	Total sample
Landscape	Mountainous	Hilly	Plain	-
Sample size	185	186	185	556
Average family Size	4.4	4.3	3.6	4.1
Arable land ownership (Mu)	4.51	4.57	2.95	4.01
Age of household head (Years)	48	53	54	52
Proportion of female household head	0.11	0.03	0.08	0.07
Educational years of household head (Years)	5.92	6.33	7.02	6.42
Net income per capita (CNY per year)	19964	22208	30793	24318

Source: Author's own field survey

The main characteristics of our sampled households are summarized in Table 1.2. In general, households from our sample have an average family size of 4.1. Only 7% of them have female heads of household. The average age of household heads is 52, while the mean value of household head educational level is about 6.42 years. The average arable land ownership is approximately 4.01 Mu[7]. The average annual per capita net income is 24318 CNY. With respect to the three income-level zones, households from Aba are living in mountainous areas with the lowest per capita net income of 19,964 CNY per year, whereas households located in Deyang within the plain areas have the highest annual average per capital net income level of 30,793 CNY. Furthermore, for the households from Yibin in hilly areas, the average per capita net income level is 22208.2 CNY per year. Moreover, households in high-income zones have the smallest family size of 3.6 and the least arable land areas of 2.95 Mu. Their household heads have the highest average age and educational level. On the contrary, households from low-income zones have the largest family size of 4.4. They also have the youngest heads with the lowest educational level. The fraction of female in household heads of these households is higher than that of those households from other two zones.

In addition, nearly 99.3% of our surveyed households employ more than one type of energy for residential purposes. Their energy use status is listed in Figure 1.7.

---

7   1 Ha= 15 Chinese Mu

Figure 1.7 Surveyed households' energy use for residence in study regions

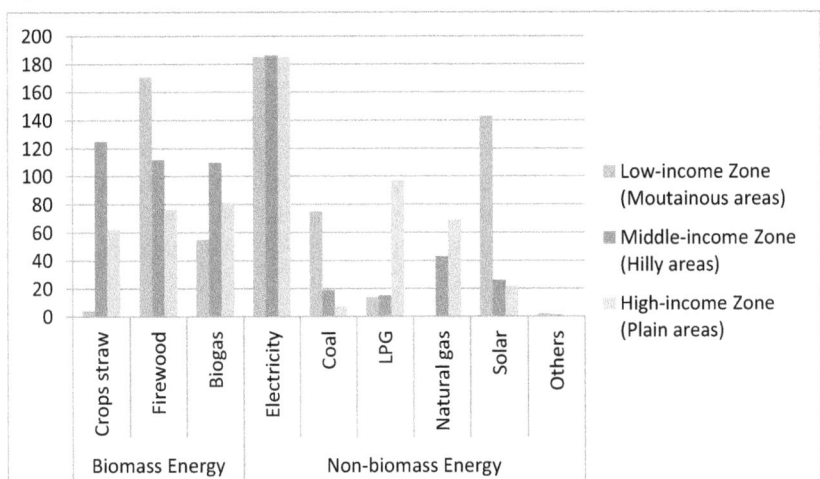

Source: Author's own field survey

Figure 1.7 shows that all the surveyed households use electricity. Households from plain and hilly areas have access to the Chinese national grid, whereas those from mountainous areas have been connected to nearby small hydropower generation stations. The most commonly used biomass energy for the sampled households is firewood, accounting for approximately 64.6% (359) of the total population. About 44.2% of the surveyed households (246) use biogas, while crop straw takes the share of 34.4% (191). This means that the biomass energy still occupies a relatively large proportion in current household energy consumption structure.

With respect to regional differences, households from different areas have different energy choices. The low-income households from the mountainous areas depend heavily on firewood (92.4%) for cooking and space heating. Solar (77%) is the major type of non-biomass energy used to heat water for showers, as the high-altitude zones usually have adequate sunshine. In addition, coal (40.5%) is an important type of commercial energy for these households to heat their houses in the cold winter. Among the medium-income households, the biomass energy takes a dominant position in their residential energy consumption. The percentage of households using crop straw, firewood, and biogas are 67.2%, 60.2%, and 59.1%, respectively. In addition to electricity, natural gas is another commonly used commercial energy source in hilly areas, having been adopted by 23.1% of the households there. With regard to the high-income households who are living in plain areas, biogas (43.8%) is the most popular type of biomass energy used for cooking. Among all the other energy alternatives, LPG has the largest user share of about 52.4% in this subsample. Compared to the other two household groups, the proportion of biomass energy consumers is smaller, while the proportions of LPG and natural gas users are larger in this richer group. As presented in Table 1.1 and Figure 1.6, households from high-income zones (plain areas) are more likely to use cleaner commercial energy with higher efficiency and quality such as biogas, natural gas, and LPG, but without abandoning the traditional solid biomass energy.

1.7 Main Contributions of the Thesis

The main contributions of this thesis are as follows:

Firstly, the issue of the impacts of biomass energy use on rural livelihoods has been given insufficient attention in the foregoing literature in China. This research will conduct a quantitative analysis using Chinese data.

Secondly, it fills the gaps from past literature and provides holistic, comprehensive, and in-depth research on the impacts of household use of biomass energy on rural livelihoods.

Thirdly, from a methodological point of view, this research advances the literature by triangulating the existing approaches to robustly analyze household energy use behaviors at the micro level.

1.8 Organization of the Thesis

This thesis aims to address the research questions proposed in Section 1.4. We adopt data collected from our household survey in Sichuan Province of China and attempt to make some contributions to the heated debates over the impacts of household biomass energy use on rural livelihoods. The structure of this research is organized as follows:

Chapter 2 develops a theoretical analysis of how household biomass energy use affects rural livelihoods based on the classical agricultural household model provided by Benjamin (1992). We also lay out the theoretical scenarios for subsequent empirical chapters in this thesis concerning the relationship between biomass energy use and livelihoods.

Chapter 3 estimates how households make decisions on cooking energy choice. This chapter also tests the main determinants of household fuel choice behaviors. Households' revealed and stated preferences for energy use are discussed based on random utility theorem and the empirical technique of choice modeling.

Chapter 4 examines the impacts of biomass collection on agricultural production from a perspective of profit maximization in order to provide insights regarding the relationship between these two activities.

Chapter 5 studies the effects of the changes in exogenous markets on household biomass energy use. Focusing on household labor allocation, we not only jointly analyze both production and consumption sides of the household decision making on the use of biomass energy, but also study how the exogenous prices affect household biomass energy use.

In Chapter 6, we draw some conclusions and try to provide some useful and feasible suggestions and policy implications based on the theoretical and empirical analysis conducted in previous chapters.

Chapter 2 Analytical Framework

2.1 An Agricultural Household Model

As rural households play the double role of supplier-consumer in domestic biomass energy utilization, the interactions between their decisions on what types of energy to use, how to allocate their time endowments into agricultural production, biomass collection and off-farm employment, and how to adjust their energy consumption decisions in response to the changes in food prices, commercial energy, and other marketing goods identifies their basic livelihood strategies. This also provides motivation for this research to examine the influence of domestic biomass energy utilization on rural livelihood, jointly taking into account biomass collection and biomass energy consumption. Thus, the agricultural household model could be an appropriate tool applied for analysis in this context (Heltberg et al., 2000; Carter and Yao, 2002; Fisher et al., Chen et al., 2006).

In previous works on microeconomics, household decision-making behaviors were usually analyzed at two levels. At the intrahousehold level, the individual-based decision-making process emphasizes the impacts of the potential interactions among household members and their preference heterogeneity (Kusago and Barham, 2001), while at the household level, the preferences of household members are assumed to be unified (Nepal et al., 2005). In our study, within a household, male and female members or members of different generations indeed have different preferences on energy utilization (e.g. Women are mainly responsible for cooking, while men play the dominant role in production activities). Although different approaches such as cooperative bargaining models (Manser and Brown, 1980) and collective models (Chiappori, 1988; Browning and Chiappori, 1998) have been developed to incorporate distinctive individual preferences into the household model, the individual-based methods still lack several aspects of behavioral realism, as the extent of preference heterogeneity cannot be clearly identified and directlfy measured (Kusago and Barham, 2001). In addition, some behavioral characteristics and attributes, such as household actual performance, are only observable at the household level (Nepal et al., 2005). Meanwhile, individual decisions among household members are still affected by the characteristics of the households they belong to, especially the demographic structure and income level. Thus, due to data availability, in order to simplify our analysis, this study will focus on decision-making behaviors at the household level, basically assuming that all household members have common preferences regarding consumption and resource allocation.

Now, we start from an agricultural household model with biomass energy use. The model is adapted on the basis of the classical model provided by Benjamin (1992). It integrates biomass collection and biomass energy consumption into the intrahousehold economic activities for investigating how a household makes decisions on biomass energy utilization and how it would, in turn, influence household livelihoods.

A twice-differentiable quasi-concave household utility function can be defined as follows:

$$U_i(C_i, l_i; a_i) \tag{2.1}$$

Where subscript i indexes the individual households to which the household belongs; vector a is a set of household characteristics which can influence preferences; l is the denotation of leisure; and C is the total household consumption, the sum of the market purchased and home produced commodities. Herein, we divide C into two categories following Amacher et al. (1996), Heltberg et al. (2000), and Charles and James (2008): consumption of goods and services that requires energy inputs C_h and other marketed goods and services C_m. This can be expressed in the following equation:

$$C_i = C_{hi} + C_{mi} \tag{2.2}$$

In the context of Sichuan Province, the household goods and services, including cooking, heating, and lighting, are mainly produced with commercial energy and biomass energy:

$$C_{hi} = \Gamma(C_{ei}, C_{bi}; S_i) \tag{2.3}$$

Where C_e denotes household consumption of commercial energy (i.e. coal, electricity, LPG, etc) and C_b denotes biomass energy (crop straw and firewood, which we defined before) consumption. S is a set of factors that influence energy using efficiency (i.e. possess of improved stove, and cooking or heating habits, etc).

The agricultural production of the household is assumed to be continuous and monotonic in L_{ai}, twice differentiable, and strongly concave. It is represented by the function:

$$q_{ai} = F_{ai}(L_{ai}; B_i) \text{ with } F'_{ai} > 0, F''_{ai} < 0 \tag{2.4}$$

Where B is a set of all inputs except labor (i.e. land, water, and all the other inputs) which is assumed to be exogenous.

Similarly, we assume that the labor supplied to biomass collection is L_{bi} (≥ 0) and define the biomass collection function as:

$$q_{bi} = F_{bi}(L_{bi}; Z_i) \text{ with } F'_{bi} > 0, F''_{bi} < 0 \tag{2.5}$$

Where Z is an exogenous vector of characteristics pertaining to the accessibility and availability of biomass resources such as the distance from the forest or the field to the house, the transportation cost, and the stock of biomass resources.

We also assume that a household has fixed time endowment $T(a)_i$, which can be divided into four non-overlapping livelihood activities: working on farm for production profits (L_{ai}), working for biomass collection (L_{bi}), working off-farm for wage (L_{oi}), and leisure for welfare maximization (l_i). Hence, we have:

$$T(a)_i = L_{ai} + L_{bi} + L_{oi} + l_i \tag{2.6}$$

Then, after normalizing prices for all goods by agricultural output price to simplify our analysis, we can get to know the optimization problem for the household is:

$$MaxU_i[C_{mi} + \Gamma(C_{ei}, C_{bi}; S_i), T(a)_i - L_{ai} - L_{bi} - L_{oi}; a_i]$$
$$\text{w.r.t. } C_i, L_{ai}, L_{bi}, L_{oi}, l_i, \text{ s.t. } C_i = \pi_i + w_i L_{oi} + E_i \tag{2.7}$$

Where the prices of all goods are normalized by the price of agricultural products; π is the profits from production activities; E is the exogenous income of the household which includes remittances, transfers, and all the other real non-labor income; and w is the wage rate for labor.

Problem (2.7) is the starting point model. Before going into the deeper analysis of household biomass energy utilization behaviors, we firstly concentrate on household preferences for different types of energy (i.e. a in the agricultural household model), especially the biomass energy, and try to find out the determinants.

2.2 Household Energy Choice and Its Determinants

In the first step, we want to investigate the household choice behaviors on biomass energy from the perspective of household preferences for different types of energy and to find the factors that can affect these behaviors. Energy Economics proposes two main theoretical frameworks developed to analyze the factors influencing the choices and behaviors of the energy utilization by rural households.

The first one is the energy ladder model, which was previously the dominant model applied to study the energy choice behaviors of the households in developing countries (Leach, 1987; 1992). It arranges an array of fuels from the worst to the best in terms of cost, cleanliness, convenience, technological sophistication, and so on (Hosier and Dowd, 1987; Smith et al., 1994; Sudhakara and Reddy, 1995; Arnold et al., 2006; Van der Kroon et al., 2013; Gosens et al., 2013). As a type of modern energy with high efficiency and little pollution, electricity is supposed to be at the top of the "ladder," whereas the traditional solid biomass energy (such as crops straw, firewood, and animal dung) is placed at the bottom due to its low-efficient utilization and negative impacts on livelihoods (Smith et al., 1994; Bruce et al., 2002). The basic assumption for this model is that, with the improvement of the economic status, the rural households can move up along the ladder to the 'better' energy carriers (Sudhakara and Reddy, 1995; Masera et al., 2000; Bruce et al., 2002). Thus, income is the most important determinant of household energy choice (Leach, 1987). This indicates that a crucial reason for the slow energy transition process in some backward rural areas is that the poor are highly likely to be trapped by the high prices of the modern high-quality fuels and then must depend on low-quality energy for living (Gosens et al., 2013). Specifically, the main contributions and limits of this theoretical model are as follows:

Contributions:

- The energy ladder model is derived and established on the basis of the microeconomic theory of rational choice (Masera et al., 2000). That is to say, the household could be assumed to behave as a neoclassical consumer who will pursue maximum utility in this model (Leach, 1987; 1992). Therefore, it can be helpful to test the effect of income level on energy choice.
- The main advantage of the energy ladder model is that it is able to capture the strong dependency of household energy choice on income (Van der Horst and Hovorka, 2008; Sovacool, 2011).

Limits:

- The fuel adoption is assumed to be a linear progress in the energy ladder model, implying that climbing up the ladder will be accompanied with the corresponding abandonment of fuel at the lower level (Kowsari and Zerriffi, 2011). However, this assumption is usually inconsistent with reality (Masera et al., 2000).
- This approach, to some extent, overemphasizes the income effect, as it neglects the impacts of other factors, which can also significantly affect the household energy choice at the same time (Leach 1988; Karekezi and Majoro, 2002; Heltberg et al., 2004, 2005; Ouedraogo, 2006).

The second theoretical model that has been widely used in the analysis of household energy choice behaviors is the energy stacking model. It thinks that households may consume a combination of several types of fuels, which simultaneously contains traditional and modern ones at different levels along the energy ladder (Van der Kroon et al., 2013). That is to say, the energy stacking model also relies heavily on the universal hierarchical order of different fuel choices and services that has been assumed in the energy ladder model (Takama et al., 2012), and the household energy choice may also tend towards the high quality energy with the growth in their socio-economic values. Nonetheless, the multiple fuel use patterns of household are determined by the complex interactions among various factors such as income (or wealth), local food and cooking habits, local tradition and institution, ethnicity, and food taste preferences (Masera et al., 2000; Karekezi and Majoro, 2002; Heltberg et al., 2005; Ouedraogo, 2006; Takama et al., 2012).

Contributions:

- This theory has shown that fuel switching occurs partially among the majority of the households, based on a growing number of empirical evidence (Israel 2002; Heltberg et al., 2004, 2005; Gupta and Köhlin, 2006; Ouedraogo, 2006; Farsi et al., 2007; Takama et al., 2012). As household economic status increases, changes in energy choice can be regarded as a process of "accumulation of energy options" rather than a linear progress (Masera et al., 2000).
- Unlike the energy ladder model, the energy stacking framework considers a wider range of factors affecting household energy choice, although income is still one of the most important determinants of fuel choice (Israel 2002; Karekezi and Majoro, 2002; Heltberg et al., 2004, 2005; Ouedraogo, 2006; Gupta and Köhlin, 2006; Farsi et al., 2007; Takama et al., 2012).
- Energy stacking is a ladder of energy demand rather than fuel preferences, which means that the energy choice of the household is driven by the energy demand of the household, which in turn is elicited by the services that energy provides (Foley, 1995; Masera et al., 2000).

Limitations of the model:

- This model relies heavily on a universal hierarchy of different types of energy and still places too much emphasis on the role of incomes in energy choice (Kowsari and Zerriffi, 2011).
- It lacks consideration for sustainability because of the dimension systems which can be used to specify the contributions of the different types of energy to sustainability has not been set up yet (Gosen et al., 2013).

Furthermore, recent years have brought the development of another important concept of energy leapfrogging to explain household energy choice behaviors. This term refers to the household energy transition pathways which bypass the conventional energy and directly skip to the modern and clean energy technologies (Murphy 2001). Nevertheless, in practice, energy leapfrogging is often deemed to be misleading (Murphy, 2001; Wolfram, 2014; Guta et al., 2015), as it cannot be achieved without simultaneous leapfrogging in local economies and institutions (Han et al., 2008).

Though the above approaches have tried to explain the nature of the energy transition from the traditional biomass energy to the modern and clean energy carriers with the increasing wealth, and provided analytical tools to study the biomass energy adoption of rural households, they are mainly tested using the data on the observed choice or consumption behaviors in the existing literatures (Hosier and Dowd, 1987; Sudhakara and Reddy, 1995; Heltberg et al., 2004, 2005; Ouedraogo, 2006; Gupta and Köhlin, 2006; Farsi et al., 2007). In many cases, the data of some important potential influencing factors cannot be directly observed and effectively collected. Hence, this research will try to fill the gaps and provide a holistic, systematic, and in-depth analysis on the household choice behaviors, jointly considering the revealed and stated preferences of the household. According to the real situation of Sichuan Province and our research objectives, we assume that a group of fuel alternatives faced by rural households can be ranked in order in terms of cost, cleanliness, efficiency, and technological flexibility. Electricity is at the top, whilst the traditional biomass energy (crops straw and firewood) is at the low end of the range. The process of cooking fuel switching can be characterized by household switching from solid biomass through conventional solid energy (coal) and gaseous fuels (biogas, LPG and natural gas) to advanced energy (electricity).

In particular, biomass energy, which includes traditional biomass energy and biogas, is produced by households based on biomass collection and energy preparation, whereas coal, natural gas, LPG, and electricity can only be purchased from the market. Energy-specific attributes such as energy prices (market prices of coal and electricity and the shadow prices of traditional biomass energy and biogas), smoke level, and safety should be the influencing factors responsible for household energy use choice behaviors. Additionally, the energy use choice should be household specific. It is therefore also influenced by household characteristics. Household income level is still regarded as an important factor that can affect energy use choice decision. Poorer households are expected to rely more on traditional biomass energy (Gupta and Kölin, 2006). With the increase in income level, households are more likely to choose higher quality energy. In our research, we will focus on household characteristics such as the characteristics of the decision maker, the demographic characteristics of the family, the cooking habits, and household location. The specific variables used for empirical analysis will be selected based on a review of previous works of literature.

2.3 The Impacts of Biomass Collection on Agricultural Production

In the second step, we want to shed some light on the household decision-making behaviors on the production side. Three basic assumptions were set in this part. The first is that household consumption and production decisions are non-separable. The second is that the intrahousehold economic activities on the production side are only composed of biomass collection and agricultural production. We also assume that farm households allocate their limited labor endowment to agricultural production, biomass collection, and off-farm work to generate income to support their consumption. In this context, this research will focus on how households allocate labor to biomass collection and agricultural production and try to model the relationship between these two activities.

Based on what we have defined for the basic model in Section 2.1, we now introduce the household biomass collection into our household model. In rural Sichuan, the most important types of biomass energy are crop straws collected from the farm and firewood collected from forest. Then, we define biomass resources here as crop residues and firewood.

In the previous literature using the household models, the biomass collection activities were treated in two ways. Some researchers used the non-separable household model with a focus on fuel production and consumption (Amacher et al. 1996; Mekonnen, 1997; Nyang, 1999; Joshee et al., 2000; Mishra, 2008). This model lacks information on the linkages between fuel market and other markets and ignores the impacts of other markets on household decisions. Sometimes the biomass collection was integrated into the agricultural household model by adding a separate production function (Wiedenmann, 1991; Heltberg et al., 2000; Köhlin and Parks, 2001; Fisher et al., 2005; Chen et al., 2006; Charles and James, 2008). In this approach, there is an implicit assumption that labor allocation decisions are separable and can be made independent of allocation decisions on agricultural production and biomass collection (Weaver, 1983). However, in Sichuan Province, the labor allocated between agricultural production and biomass collection cannot be distinguished by any physical indicator such as gender and age. The members of the household engaged in farm work, in most cases, are also responsible for biomass collection. They often collect firewood on their way to and from the fields or collect crop straw after harvesting and take it home. The simple aggregation of the production functions in past studies lacks the information on the internal relationship between agricultural production and biomass collection. Rural households usually rely on the market to provide signals through the price system to choose the proportions of available labor inputs that should be allocated to each activity (Debertin, 2012). In other words, the labor allocation should on the basis of the decisions regarding these two activities. Therefore, the above assumption will not be held in this study. Under this circumstance, a multiple output production function will be considered (Weaver, 1983).

The multiple output production function that embodies the behavioral relationship as well as technical relationship based on the single-input production functions (2.4) and (2.5) is defined as:

$$f_i(q_{ai}, q_{bi}) = g_i(L_i; B_i, Z_i); \quad L_i = L_{ai} + L_{bi} \tag{2.8}$$

Where the function $f(\cdot)$ is concave in q_{ai} and q_{bi}. This shows the behavioral relationship that defines the transformation curve for the agricultural products and collected biomass (Debertin, 2012). The function $g(\cdot)$ reflects the technical relationship that specifies the possible combinations of the output q_{ai} and q_{bi} produced from the mix of labor inputs L_{ai} and L_{bi} (Debertin, 2012), and it may be concave in L_i (the total labor input for intrahousehold production activities). Using the implicit function theorem, we can write:

$$q_{ai} = F_{ai}[L_i - F_{bi}^{-1}(q_{bi})] = h_i(q_{bi}, L_i; B_i, Z_i) \tag{2.9}$$

And we can also obtain:

$$L_i = F_{ai}^{-1}(q_{ai}) + F_{bi}^{-1}(q_{bi}) \tag{2.10}$$

The total differentiation of (2.10) with respect to q_{ai} and q_{bi} yields:

$$dL_i = \frac{dF_{ai}^{-1}(q_{ai})}{dq_{ai}}dq_{ai} + \frac{dF_{bi}^{-1}(q_{bi})}{dq_{bi}}dq_{bi} = (1/MPL_{ai})dq_{ai} + (1/MPL_{bi})dq_{bi} \tag{2.11}$$

Assuming that L_i is invariable, therefore we have:

$$dL_i = (1/MPL_{ai})dq_{ai} + (1/MPL_{bi})dq_{bi} = 0 \Rightarrow \frac{dq_{ai}}{dq_{bi}} = -\frac{MPL_{ai}}{MPL_{bi}} \tag{2.12}$$

The equation (2.12) gives the behavioral relationship between agricultural production and biomass collection. The expression dq_{ai}/dq_{bi} represents the slope of the product transformation curve at a particular point. It is the rate of product transformation of biomass collection for agricultural production ($RPT_{q_{ai}q_{bi}}$) and indicates the rate at which agricultural products can be substituted for the biomass outputs as the labor input bundle is reallocated (Debertin, 2012). Along the production transformation curve, $RPT_{q_{ai}q_{bi}}$ is equal to the negative ratio of individual marginal products. According to our assumptions that $F'_a > 0$ and $F'_b > 0$, this rate is unambiguously negative. This implies that, for agricultural production and biomass collection, one must be reduced in order to obtain more of the other, given a fixed available amount of labor inputs L_i.

The above description gives a theoretical explanation for adopting the multioutput production function to investigate the relationship between the two activities. As we are not quite interested in the household biomass energy consumption behaviors in this section, we then can ignore the utility function and concentrate on the household's objective to maximize its real income from agricultural production, biomass collection, and off-farm work. Then the basic problem in expression (2.7) collapses to a profit maximization problem:

$$Max\pi_i = g_i(L_i; B_i, Z_i) - w_i^*(L_{ai} + L_{bi}) + w_i L_{oi} + E_i$$
$$= f_i(q_{ai}, q_{bi}) - w_i^*(L_{ai} + L_{bi}) + w_i(T_i - L_{ai} - L_{bi} - l_i) + E_i \quad w.r.t. \ L_{oi}, L_{ai}, L_{bi} \quad (2.13)$$

With slight modification, the first-order conditions for the household labor allocation are obtained as:

$$\frac{\partial \pi_i}{\partial L_{oi}} = w_i \quad (2.14)$$

$$\frac{\partial \pi_i}{\partial L_{ai}} = \frac{\partial q_{ai}}{\partial L_{ai}} - w_i^* = 0 \Rightarrow MPL_{ai} = w_i^* \quad (2.15)$$

$$\frac{\partial \pi_i}{\partial L_{bi}} = \frac{\partial q_{bi}}{\partial L_{bi}} - w_i^* = 0 \Rightarrow MPL_{bi} = w_i^* \quad (2.16)$$

The conditions from (2.14) to (2.16) imply that the optimum of labor allocation between agricultural production and biomass collection will occur at the points at which the marginal output equals the shadow wage of household labor. Then we can solve the household profit maximization problem and obtain the reduced form of household optimal labor allocation functions as follows:

$$L_{ni}^* = L_i^*(w_i^*, w_i, T(a)_i, E_i, B_i, Z_i) \ (n = a, b, o) \quad (2.17)$$

The labor allocated to off-farm work (o), agricultural production (a), and biomass collection (b) can be expressed as a function of market wage rate, shadow wage rate, household time endowment, non-labor income, inputs and services for agricultural production, and the factors affecting biomass collection. However, the shadow wage rate of household labor is endogenous and mainly determined within the household. As it is a function of household characteristics affecting household preferences and choices (Strauss, 1986), we can transform equation (2.17) to:

$$L_{ni}^* = L_i^*(w_i, T(a)_i, E_i, B_i, Z_i, a_i) \, (n = a, b, o) \tag{2.18}$$

Now, following (2.8) to (2.18), we have already known the relationship between agricultural production and biomass collection from theoretical perspectives. In the empirical part, we will estimate a multi-output profit function to examine household production decisions towards these two activities and clarify the relationship between them based on what we have discussed above.

2.4 The Impacts of Household Biomass Energy Utilization Behaviors on Rural Livelihood

In the case of Sichuan Province, to meet its energy demand, each household has to decide whether to use biomass energy and which types of energy it will use. Some rural households only use cleaner commercial energy such as electricity, LPG, and natural gas, while the others use a combination of commercial energy and biomass energy. Compared with the households who merely use commercial energy, the members of the households using biomass energy have to spend more time on biomass collection and could be constrained from participating in other income-generating activities such as agricultural production and off-farm work. In other words, if a household decides to use biomass energy, its members have to allocate their limited time endowment onto biomass collection, leisure, and other activities.

Under this circumstance, we assume that a typical agricultural household plays the double role of consumer-supplier in the markets of agricultural products and biomass energy. It jointly makes its decision on the production and consumption of agricultural products and biomass energy. On one hand, it has to allocate its limited labor resources onto three different activities, namely agricultural production, biomass collection, and off-farm employment. On the other hand, it maximizes its utility from consumption of commercial energy, biomass energy, other purchased goods and services, and leisure time subject to a number of constraints. Particularly, the household budget limitation is depicted by the full income constraint for all purchased goods and services, which consists of the labor returns from these three activities and the exogenous income. In this section, we will further develop the agricultural household model.

2.4.1 Impacts of Biomass Energy Utilization on Rural Livelihood in a Separable Agricultural Household Model

A separable agricultural household model with biomass energy is considered. According to the separation property given by Benjamin (1992), household behaves in a recursive process. It firstly maximizes its profits from the multi-output production function without any consideration of its consumption or leisure preferences (Benjamin, 1992; Skoufias, 1994):

$$Max \pi_i = g_i(L_i; B_i, Z_i) - w_i^* L_i = f_i(q_{ai}, q_{bi}) - w_i^* (L_{ai} + L_{bi}) \quad w.r.t.\ L_i, L_{ai}, L_{bi} \quad (2.19)$$

Where w_i^* denotes the shadow wage of household labor.

The first-order conditions are as follows:

$$\frac{\partial \pi_i}{\partial L_i} = \frac{\partial g_i(L_i; B_i, Z_i)}{\partial L_i} - w_i = 0 \Rightarrow MPL_i = w_i \quad (2.20)$$

$$\frac{\partial \pi_i}{\partial L_{ai}} = \frac{\partial q_{ai}}{\partial L_{ai}} - w_i = 0 \Rightarrow MPL_{ai} = w_i \quad (2.21)$$

$$\frac{\partial \pi_i}{\partial L_{bi}} = \frac{\partial q_{bi}}{\partial L_{bi}} - w_i = 0 \Rightarrow MPL_{bi} = w_i \quad (2.22)$$

Based on the conditions from (2.20) to (2.22), the new equilibrium of household labor allocation is determined by:

$$MPL_i = MPL_{ai} = MPL_{bi} = w_i \quad (2.23)$$

The equation (2.23) indicates that the optimum of household labor allocation will occur at the point at which the marginal productivity of the labor allocated to intrahousehold production activities is equalized to the marginal productivity of the farm labor and labor in biomass collection. It also demonstrates that household labor allocation on agricultural production and biomass collection is determined by shadow wage.

Thus, the solution to the maximization problem yields the household labor allocation functions for these two activities:

$$\left. \begin{array}{c} L_{ai}^* \\ L_{bi}^* \end{array} \right\} = L_i(q_{ai}^*, q_{bi}^*, L_i^*) = L_i(w_i; B_i, Z_i) \quad (2.24)$$

Figure 2.1 gives a visual demonstration for the equilibrium conditions of household labor allocation between agricultural production and biomass collection. In the plane $q_{ai}q_{bi}$, each production-possibility curve represents production tradeoffs of the household given the fixed level of labor inputs L_i^*. It shows the various combinations of amounts of two products that the household can produce. In accordance with the relationship in (2.12), the shape of the production-possibility curve should be concave to origin, and with an increase in L_i^*, the curve will move outward driven by the corresponding increase in outputs. Condition (2.23) also reveals that, at equilibrium, the household will choose to produce the combination at the point at which the slope of the production-possibility curve is equal to negative one. If the wage rate increases, the household will accordingly increase the labor inputs for both of agricultural production and biomass collection, and then, the equilibrium point will move along the straight line from D_1 to D_2. Conversely, if there is a decrease in wage rate, the production tradeoff of the household will shift from D_3 to D_4.

Figure 2.1 Pathway of the equilibrium on the production-possibility curve of the household with decrease (Left) and increase (Right) in wage rate

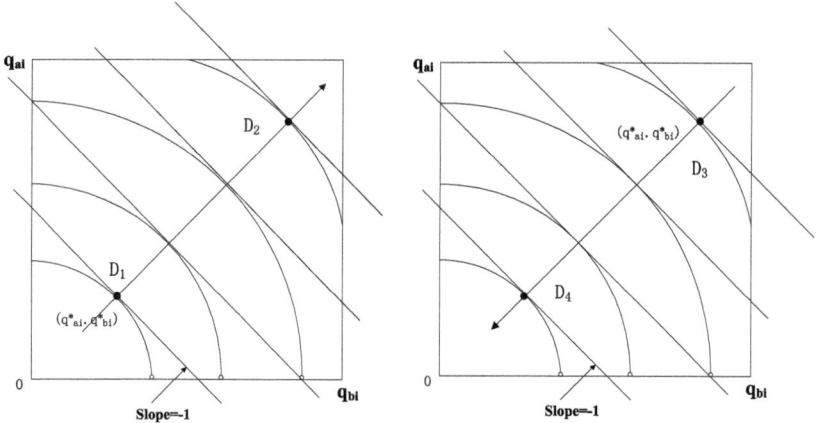

Source: Author's depiction

Furthermore, based on the maximum profits, the household then decides how much leisure to consume and how much labor to supply off-farm as the solution to the basic model in (2.7):

$$l_i^* = l_i(w_i, M_i^*; a_i, B_i, Z_i, S_i) \tag{2.25}$$

$$L_{oi}^* = T(a)_i - L_i^* - l_i^* \tag{2.26}$$

Where $M_i^* = \pi_i^* + w_i T(a)_i + E_i$ is the optimal budget of the household. Therefore, the household consumption decision will be made as:

$$\left.\begin{array}{l} C_{ei}^* \\ C_{bi}^* \\ C_{mi}^* \end{array}\right\} = C_i(w_i, M_i^*; a_i, B_i, Z_i, S_i) \tag{2.27}$$

Further illustration of the equilibrium is provided in Figures A.1 to A.3. Figure A.1 gives the household optimal labor allocation (V) for intrahousehold economic activities, which consist of agricultural production and biomass collection under the separation assumption according to condition (2.24) in a three-dimensional space. With the changes in the market wage rate, the corresponding changes in equilibrium from V to V' are shown in Figures A.2 and A.4. There is a negative relationship between labor allocated to intrahousehold economic activities and the market wage rate. When the market rate increases (or decreases), the amount of labor engaging in intrahousehold economic activities will reduce (or increase) accordingly.

2.4.2 Impacts of Biomass Energy Utilization on Rural Livelihood in a Non-separable Agricultural Household Model

It is well recognized that, in developing countries, market imperfections are pervasive in rural areas. As in the case of Sichuan, there may be an exogenously imposed binding constraint on the labor market in rural areas. There are several plausible reasons for this in the Chinese context: the relatively low educational level of rural households for getting a job off-farm, the high transaction cost of the inconvenient transportation system in some remote areas, the fear of losing the land use rights of the household members working off-farm (Wang et al., 2011; Jia, 2014), etc.. With regard to the energy market in rural area, the domestically produced biomass energy is almost not traded. Rural households face two choices, and they choose either to purchase different kinds of commercial energy from the market or to collect biomass by themselves based on their valuations of their own labors (Amacher et al., 1996). Under these circumstances, the separation property breaks down, and we should consider a non-separable household model.

Now, we assume that the market for commercial energy is perfect. Meanwhile, the biomass collected by households is assumed to be non-tradable; that is, the consumption is lower than supply[8]. Then, we have the constraint for household biomass energy consumption:

$$C_{bi} \leq q_{bi} \tag{2.28}$$

In particular, the labor market imperfections are introduced in the agricultural household model as a binding constraint on off-farm labor:

$$L_{oi} \leq H_i \tag{2.29}$$

In order to simplify our analysis, we also assume that the markets for all the other goods and services are perfect and normalize prices of all other goods by exogenous market price (p_i) to get a new full-income constraint for the household:

$$C_{mi} + C_{ei} = q_{ai} + w_i L_{oi} + E_i = h_i(q_{bi}, L_i; B_i, Z_i) + w_i L_{oi} + E_i \tag{2.30}$$

Then we solve the optimization problem of the household in (2.7) by establishing the Lagrangian function subject to the constraints (2.28), (2.29), and (2.30):

$$\begin{aligned} U_i^L &= U_i[C_{mi} + \Gamma(C_{ei}, C_{bi}; S_i), T(a)_i - L_{ai} - L_{bi} - L_{oi}; a_i] + \lambda_i [h_i(q_{bi}, L_i; B_i, Z_i) + w_i L_{oi} + E_i \\ &\quad - C_{mi} - C_{ei}] - \mu_i(L_{oi} - H_i) - \eta_i(C_{bi} - q_{bi}) \quad w.r.t. \ C_{bi}, C_{ei}, C_{mi}, L_{oi}, L_i, L_{ai}, L_{bi}, q_{bi} \end{aligned} \tag{2.31}$$

The Kuhn-Tucker conditions can be obtained through the following:

$$\frac{\partial U_i}{\partial \Gamma} \frac{\partial \Gamma}{\partial C_{bi}} = \eta_i \tag{2.32}$$

$$\frac{\partial U_i}{\partial C_i} = \frac{\partial U_i}{\partial C_{ei}} = \frac{\partial U_i}{\partial C_{mi}} = \lambda_i \tag{2.33}$$

$$\frac{\partial U_i}{\partial l_i} = \lambda_i w_i - \mu_i \tag{2.34}$$

$$\frac{\partial U_i}{\partial l_i} = \lambda_i \frac{\partial q_{ai}}{\partial L_i} + \eta_i \frac{\partial q_{bi}}{\partial L_i} \tag{2.35}$$

$$\frac{\partial U_i}{\partial l_i} = \lambda_i \frac{\partial q_{ai}}{\partial L_{ai}} \tag{2.36}$$

---

8 A part of the agricultural residues is used as fertilizer in most of the cases in our study region.

$$\frac{\partial U_i}{\partial l_i} = \lambda_i \left(\frac{\partial q_{ai}}{\partial q_{bi}} \frac{\partial q_{bi}}{\partial L_{bi}} + \frac{\partial q_{ai}}{\partial L_i} \frac{\partial L_i}{\partial L_{bi}} \right) + \eta_i \frac{\partial q_{bi}}{\partial L_{bi}} \tag{2.37}$$

$$\lambda_i \frac{\partial q_{ai}}{\partial q_{bi}} + \eta_i = 0 \tag{2.38}$$

Using equation (2.12) and replacing it in (2.30), thus:

$$\frac{\partial U_i}{\partial l_i} = \eta_i \frac{\partial q_{bi}}{\partial L_{bi}} \tag{2.39}$$

And we have:

$$\frac{\partial U_i}{\partial l_i} = \lambda_i \frac{\partial q_{ai}}{\partial L_{ai}} = \frac{\partial U_i}{\partial \Gamma} \frac{\partial \Gamma}{\partial C_{bi}} \frac{\partial q_{bi}}{\partial L_{bi}} = \lambda_i \frac{\partial q_{ai}}{\partial L_i} + \eta_i \frac{\partial q_{bi}}{\partial L_i} = \lambda_i w_i - \mu_i \tag{2.40}$$

As depicted in Figure 2-d, condition (2.40) reveals that the household collects biomass until the point at which the marginal utility of leisure equals the marginal utility of biomass energy in household consumption times the marginal product of biomass collection labor which, in turn, is equalized to the marginal value product of labor in agricultural production. In other words, the marginal value product of labor in biomass collection and labor working on-farm is equal to the opportunity cost of the household labor (the utility of leisure). This result is in line with the findings of Heltberg (2000) that biomass collection is determined by the opportunity cost of the time (shadow wage) of the household labor. It also states that the time is allocated among biomass collection, farm work, off-farm employment, and leisure relying on wage rate.

Under the conditions of non-separation, the shadow wage (w_i^*) determines the labor allocation of the household. However, it is determined within the household (Singh et al., 1986). The reduced form of household labor allocation functions can be derived considering the exogenous prices on consumer market (p_i):

$$\left.\begin{array}{l} L_{ai}^* \\ L_{bi}^* \end{array}\right\} = L_i'(w_i, p_i; B_i, Z_i, S_i, a_i) \tag{2.41}$$

And then yields the shadow full income of the household:

$$Y_i^* = \pi_i^* + w_i^*[T(a)_i - H_i] + w_i H_i + E_i = F_{ai}(L_{ai}^*; B_i) + F_{bi}(L_{bi}^*; Z_i) + w_i H_i + w_i^* l_i + E_i$$

Where, $\pi_i^* = F_{ai}(L_{ai}^*; B_i) + F_{bi}(L_{bi}^*; Z_i) - w_i^* L_{ai}^* - w_i^* L_{bi}^*$ \hfill (2.42)

As a consumer, the household decides the level of consumption to maximize its utility under the shadow full income constraint (de Janvry et al., 1991). This leads to a consumption system for the household as follows:

$$\left.\begin{array}{l} C^*_{ei} \\ C^*_{bi} \\ C^*_{mi} \\ l^*_i \end{array}\right\} = C'_i(w_i, p_i, Y^*_i; a_i, B_i, Z_i, S_i) \qquad (2.43)$$

As it is shown in the expressions for optimal labor allocation and leisure consumption of the household, (2.41) and (2.43), in the non-separable model, a change in any of the exogenous variables affecting the production choices of the household will influence the labor allocation and consumption decisions of the household, both directly and indirectly. The direct effects come from the changes in households' shadow profits, as in the separable model we discussed before, whereas the indirect effects occur through the changes in the shadow wage. We will analyze both of effects in the following subsections (Skoufias, 1994).

2.4.2.1 Response of Shadow Wage to the Changes in Exogenous Market Prices

We have already determined that households allocate labor to agricultural production and biomass collection to the point at which the marginal value product of these activities equals the opportunity cost of the time, that is, the shadow wage of the household labor. Therefore, how shadow wage changes in response to a change of prices in the exogenous market presents a way to analyze the indirect effects of shadow wage on household labor allocation and consumption decisions.

Suppose now that there is only one constrained market for labor, and the exogenous market price p_i changes. Let L^*_i denote the optimal labor allocated on intrahousehold activities. Following the method provided by de Janvry et al. (1991), total differentiation of the household time endowment $-L^*_i + (T(a)_i - H_i) = l^*_i$, which determines the endogenous shadow wage w^*_i, and substitution of the quantity of labor allocated to intrahousehold activities and leisure consumption derived from the equations (2.41) to (2.43) gives:

$$-\frac{\partial L_i}{\partial w^*_i} dw^*_i - \frac{\partial L_i}{\partial p_{xi}} dp_{xi} = \frac{\partial l_i}{\partial w^*_i} dw^*_i + \frac{\partial l_i}{\partial p_{xi}} dp_{xi} + \frac{\partial l_i}{\partial Y^*_i}\left[\frac{\partial \pi^*_i}{\partial w^*_i} + (T(a)_i - H_i)\right] dw^*_i + \frac{\partial l_i}{\partial Y^*_i}\left(\frac{\partial \pi^*_i}{\partial p_{xi}}\right) dp_{xi}$$

After rearranging the equation, we can get the following new equation:

$$-\left[E_i^* + \sigma_i^* \gamma_i + \theta_i^* \gamma_i S_{li}\right]\frac{dw_i^*}{w_i^*} = \left[E_{xi}^* + \sigma_{xi}^* \gamma_i + \theta_i^* \gamma_i S_{xi}\right]\frac{dp_{xi}}{p_{xi}} \tag{2.44}$$

Where

E_i^* and E_{pi}^* are the direct and cross-price elasticity of labor supply to intra-household activities.

σ_i^* and σ_{pi}^* are the direct and cross-price elasticity of leisure consumption.

θ_i^* is the full income elasticity of leisure consumption.

γ_i is the ratio l_i^*/L_i^*, $\gamma_i > 0$.

$S_{li} = \dfrac{w_i^* l_i^*}{Y_i^*}$ and $S_{xi} = \dfrac{C_{xi} p_{xi}}{Y_i^*}$ (or $\dfrac{w_i H_i}{Y_i^*}$ in the case of a change in market wage rate) are share parameters in shadow income.

Then, we can figure out the elasticity of the endogenous shadow wage w_i^* with respect to the exogenous commercial energy prices p_i:

$$E(w_i^* / p_{xi}) = -\frac{E_{xi}^* + \gamma_i(\sigma_{xi}^* + \theta_i^* S_{xi})}{E_i^* + \gamma_i(\sigma_i^* + \theta_i^* S_{li})} \tag{2.45}$$

The elasticity in (2.45) reflects the response of the shadow wage to the changes in exogenous market price. The numerator demonstrates the disequilibrium created by a change in the exogenous price p_i on the imperfect labor market. The first term E_{xi}^* is the labor allocation change, while the second term shows the change in leisure consumption coming from the cross-price effect σ_{xi}^*. We can expect that labor supply to intrahousehold production activities responds negatively to an increase in exogenous price and that leisure consumption has a positive cross-price elasticity with respect to other marketed goods. Therefore, the sign of the numerator is ambiguous, because the magnitudes of these two terms are unknown. Analogously, the expression in the denominator reflects the disequilibrium caused by the changes in endogenous shadow wage. The direct price elasticity of labor supply is expected to be positive. The term in parentheses demonstrates the elasticity of leisure consumption, which will be positive if the direct price effect $|\sigma_i^*|$ is less than the full income effect times the share parameter $\theta_i^* S_{li}$ due to the shadow wage increases. Hence, the sign of the elasticity (2.45) cannot be derived unambiguously.

2.4.2.2 The Effects of Exogenous Market Prices on Household Biomass Energy Use

The household decisions on biomass energy consumption can also be viewed as the response of its demand for biomass energy on the markets that exit. The global biomass energy consumption elasticity with constrained labor market can be directly obtained by differentiation of (2.43):

$$E(C_{bi}/p_{xi})^G = E(C_{bi}/p_{xi})^H + E(w_i^*/p_{xi})[E(C_{bi}/w_i^*)^H + \theta_{bi}S_{li}] \tag{2.46}$$

Where $E(C_{bi}^*/p_{xi})^H = \sigma_{bxi}^* + \theta_{bi}^* s_{bi}$ is the elasticity of biomass energy consumption in a separable household model with all markets, which includes the cross-price elasticity of biomass energy consumption σ_{bxi}^* and the income effect on biomass energy consumption $\theta_{bi}^* s_{bi}$ specific to the household model. $E(C_{bi}^*/p_{xi})^H$ can be viewed as the direct effect of the exogenous price on biomass energy consumption, while the indirect effect through the changes in internal price (shadow wage) is expressed as the second term on the right of (2.46). Since the sign of (2.45) is unknown, the sign of the global elasticity thus cannot be unambiguously determined.

In accordance with the theoretical framework, which has been stated above, this study will firstly test for separability using the data collected from Sichuan Province. After dividing the sampled households into two groups (i.e. separable and non-separable groups), we will analyze their behaviors respectively corresponding to the separable and non-separable household models. For the non-separable households, we will firstly calculate their shadow wage rates and then include them in the household demand system and labor supply system to jointly estimate household biomass energy use behaviors. For the separable households, we will directly estimate the demand system to capture the household biomass energy use.

2.5 Conclusion

To sum up, the agricultural household model developed in our research indicates that household biomass energy-use behaviors are influenced by various factors. As consumers, households make decisions on which type of energy to be used for cooking based on energy-specific and household-specific variables. Among them, energy-specific variables include the economic attributes (such as energy prices, device use costs, and device maintenance costs) as well as the physical attributes (such as smokiness level and safety) of the energy alternatives, while household-specific variables refer to the household socio-economic characteristics affecting household energy choice behaviors (including income level, demographic structure, household location, and other characteristics). As producers, households have to decide how to allocate their limited labor force to biomass collection and agricultural production based on some influencing factors such as market wage rates of household members, time endowment, and non-labor income. As biomass collection competes with agricultural production for labor resources, the shadow wage rate of on-farm labor and its determinants are also important affecting factors for the relationship between these two activities. Simultaneously considering both the consumption and production sides in the non-seperable model, the effects of household biomass energy use on livelihoods are from the changes in households' shadow profits, as in the separable model, whereas the indirect effects are derived from the changes in the shadow wage. Therefore, the variables affecting biomass energy consumption such as shadow prices of biomass energy and prices of commercial energy and household income level should also be examined in this research.

Chapter 3 Household Biomass Energy Choice for Cooking in Energy Transition and Its Impacts on Rural Livelihoods

3.1 Introduction

In Sichuan Province, although the central and local governments have considered the rural energy construction as an important strategy to enhance the rural livelihoods and made great efforts to adjust the rural energy using patterns, the current energy consumption at the household level still depends heavily on traditional biomass energy such as crop straw and firewood. The long-term reliance on solid traditional biomass energy can be attributed to a rather slow energy transition process switching toward modern fuels (Gan and Yu, 2008; Démurger and Fournier, 2011). Today, with the increasing pressure imposed by the stark conflict between rural economic growth and environmental protection, the energy transition is becoming more and more urgent. Notwithstanding the fact that the Chinese government has been explicit about its objective to elicit household motivation of using high-quality energy by vigorously promoting the construction of biogas system and the electrification in rural Sichuan, the progress of energy transition is still slow there and needs to be sped up (Peng et al., 2010). More importantly, policy-making efforts in recent years have focused on encouraging households to make fuel substitutions.

Many previous studies (Reddy, 1995; An et al., 2002; IIED and ESPA, 2010; Ahmad and Puppim de Oliveira, 2015) demonstrated that the way energy transition takes place will determine its impacts on rural household livelihoods. Hence, the main purpose of this study is to investigate the transition pathways of household energy choices, with particular concern for cooking to determine ways to propel the fuel switching from traditional biomass energy to modern cleaner energy at the household level. The focus of this study is on the biomass energy, because it is the main type of energy used for cooking in rural Sichuan, and the existing evidence regarding household preferences for it are usually insufficient to give a clear picture of the current situation due to the lack of the market.

Based on what has been discussed in Section 2.2, we assume that an array of energy choices faced by the household can be ranked from the worst to the best in terms of cleanliness, use efficiency and technological sophistication. Electricity is at the top of the list, while the traditional solid biomass energy such as crops straw and firewood is at the low end of the range. In order to better understand the energy choice behaviors of the households in energy transition, we want to clarify how households make their decision regarding modern cleaner fuels and examine the determinants of these choice behaviors. The basic hypothesis of this study is that, with the increase in socioeconomic standing of the households, their energy use choices will move up from traditional biomass energy to the energy carriers at higher levels.

The structure of this chapter is as follows: In Section 3.2, related literatures are reviewed. Section 3.3 describes the general information about cooking energy consumption status in our study region. The research methodology, which is used to address our research questions and the model specification for econometric analysis, will be given in Section 3.4. Section 3.5 presents the empirical analysis using our models, and Section 3.6 will conclude the main findings of our research.

3.2 Literature Review on Economics of Energy Choices

In recent years, a large number of studies have tried to clarify the impacts of household energy use patterns on rural livelihoods (Gupta, 2003; Cherni et al., 2007; Byrne et al., 2007; Cherni and Hill, 2009; Lee et al., 2015; Biggs et al., 2015). These studies have shown that energy transition can change rural livelihoods in various ways. In accordance with the viewpoints of Chamdimba (2009), energy use is identified as two competing paradigms in rural areas of developing countries. Poor households living in geographically remote areas depend heavily on traditional biomass for most of their basic livelihood activities such as cooking and heating. Conventional burning of biomass with low thermal efficiency usually causes waste of resource (Chen et al., 2006). Additionally, it negatively impacts human health through indoor air pollution, and causes great damage to the environment and ecological system such as through deforestation, land degradation, biodiversity decrease, and soil erosion accompanying the increase in GHG emissions (Fan et al., 2011). On the other hand, in some well-developed rural areas, household demands for new efficient and environmentally friendly biofuels are gradually increasing (Chamdimba, 2009). Adopting modern fuels could bring positive effects on rural livelihoods by improving quality of life and enhancing environmental protection (Gosen et al., 2013). Although the use of modern energy sources at the household level is expected to ensure livelihood sustainability, the understanding of household energy use in energy transition and its policy implications for livelihoods are still limited. Furthermore, little attention has been paid to this issue in rural China. Therefore, this chapter uses

Chinese data to investigate household choice on different fuels, especially on biomass energy. It aims at proposing some meaningful suggestions for future energy policy design to improve rural livelihoods.

Previous studies related to household choice behaviors have mainly focused on fuel use patterns and determinants of fuel shifts in different developing economies (Ahmad and Puppim de Oliveira, 2015; Lee et al., 2015). Most of them are based on empirical analysis using econometric methods (Hosier and Dowd, 1987; Reddy, 1995; Farsi et al., 2007; Wambua, 2011; Mensah and Adu, 2013; Suliman, 2013; Ahmad and Puppim de Oliveira, 2015). Their results confirm the hypothesis that, the step on the energy ladder to which a household climbs, depends mainly on its income level. In other words, income is important in determining household fuel choice behaviors. Households with higher income normally decide to choose better-quality energy sources. Meanwhile, some recent studies have pointed out that many other factors, such as fuel characteristics, social and demographic factors, and cooking habits also affect household fuel choices (Ouedraogo, 2006; Farsi, et al., 2007; Mekonnen and Köhlin, 2008; Akpalu et al., 2011; Kwakwa et al., 2013; Mwaura et al., 2014; Ahmad and Puppim de Oliveira, 2015). Particularly, fuel price or cost, educational level of household head, family size, access to energy supplies, and regional location have attracted considerable attention in these studies. Research in China has also suggested such a range of determinants and illustrated the impacts of household energy choice on the ecological system, environment, and human health (An et al., 2002; Peng et al., 2010; Yu et al., 2012; Qu et al., 2013). The multiple fuel use patterns of households support the assumption of the energy stacking model in which households move up the energy ladder without corresponding abandonment of the energy sources at lower levels (Kowsari and Zerriffi, 2011). Among other points that cannot be ignored, the attributes of stoves associated with particular fuels including stove price, usage cost, indoor smoke emission level, and safety have been verified to have significant impacts on household cooking fuel choice (Takama et al., 2011; 2012). Nevertheless, only a few studies jointly consider the household-specific and energy-specific attributes. In most cases, household-specific factors are fixed in the short term, whereas energy-specific factors vary according to the changes in the availability of new energy products and in households' understanding of available alternatives (Takama et al., 2012). Energy-specific factors should be involved in the analysis to complement household-specific factors and to help find out market barriers for household energy transition and desirable fuel product characteristics. In addition, revealed preference data are widely applied to estimate household energy choice models in the foregoing literatures. However, the dynamic energy switching behaviors in conjunction with changes in exogenous markets cannot be predicted by stationary choice probabilities (Ben-Akiva and Morikawa, 1990). Consequently, studies using data of actual choice alone may not provide valuable knowledge for the practical design of policies targeting the promotion of the use of non-marketed energy sources or those could create public benefits (e.g. biogas). It should make clear how households intend to switch their current fuel choices in response to a planned or potential change in the rural energy system. Thus, this chapter aims to examine both household-specific and energy-specific determinants of household energy choice and provide empirical evidence on energy transition in rural China from the perspective of household revealed and stated preferences.

3.3 Household Cooking Energy Use in Study Region

According to the statistics of Sichuan Province and the data collected in our field survey, a total of 7 kinds of cooking energy sources, namely crop straw, firewood, biogas, electricity, coal, LPG, and natural gas, are currently used by the sampled households. In terms of cost, cleanliness, efficiency, and technological flexibility, the 7 cooking fuels are ranked from the worst to the best. As shown in Figure 3.1, electricity is at the top of the ladder, whilst the traditional solid biomass energy sources, such as crop straw and firewood, are at the low end of the range. In other words, electricity could be regarded as the cleanest and the most efficient cooking energy source, whereas traditional biomass energy is the most inferior.

Figure 3.1 Cooking fuel ladder in Sichuan Province

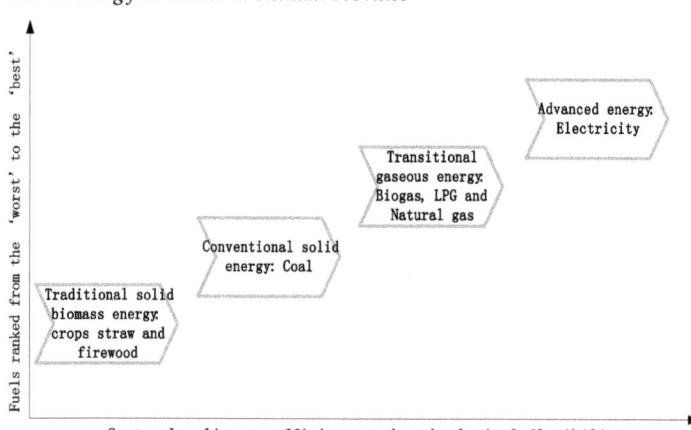

Source: adapted from WHO (2006) and Wambua (2011)

Figures 3.2 and 3.3 provide general information about household energy use for cooking (including cooking pig fodder and heating water for drinking) in our study sample. According to Figure 3.2, 528 (95%) households employ more than one type of cooking fuel at the same time. 213 households use two types of cooking fuels, accounting for about 384% of the total, whereas 34% (188) of all the households use three types of fuels simultaneously. Furthermore, a total of 127 (23%) households apply more than three types of fuel for cooking.

Figure 3.2 Cooking energy use pattern of surveyed households

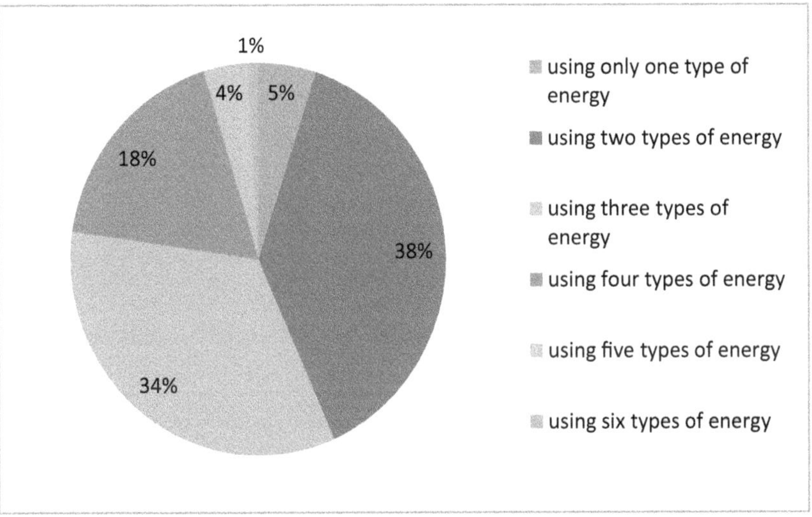

- using only one type of energy: 38%
- using two types of energy: 34%
- using three types of energy: 18%
- using four types of energy: 5%
- using five types of energy: 4%
- using six types of energy: 1%

Source: Author's own field survey

Figure 3.3 further demonstrates the household cooking fuel use in three different areas. To be more specific, electricity is the most widely used energy source. In total, 510 households choose to use electricity, accounting for 91.7% of the whole sample. There are some households that still cook without electricity because of the existing technological, organizational, and environmental problems, including grid connection difficulties, efficiency of electricity access, electrical waste, and a lack of improved discipline and coordination (Meisen and Cavino, 2007). On the contrary, coal is used by only 50 households, occupying a mere 9% of the total, due to the impacts of the local policy which has formulated that households are forbidden to use coal in some designated areas to reduce the environmental pollution caused by CO_2 emission. Biomass energy, including crop straw, firewood, and biogas, plays a vital role in cooking energy consumption structure.

Of the 556 surveyed households, 468 (84.2%) households use biomass energy. In particular, 432 (77.7%) households still adopt traditional solid biomass energy (crops straw and firewood) for cooking, while 243 (43.7%) households produce biogas. Furthermore, LPG and natural gas are employed by 120 (21.6%) and 105 (18.9%) households in our sample, respectively.

Figure 3.3 Energy use for cooking of surveyed households

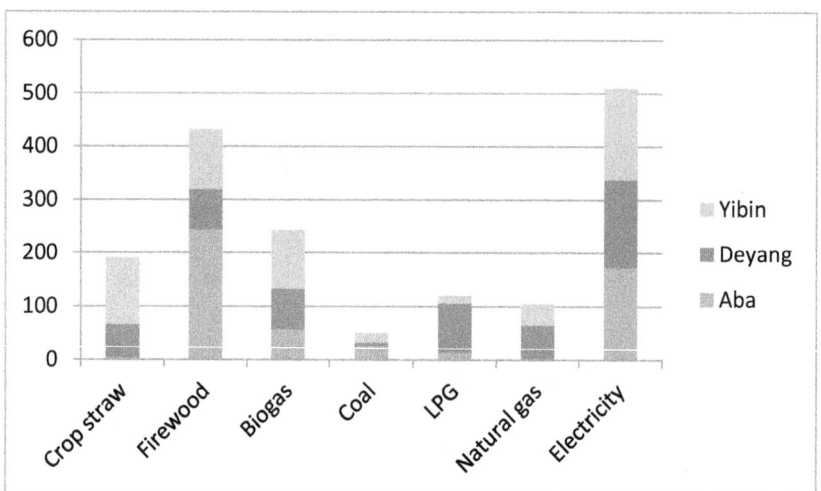

Source: Author's own field survey

In addition, households from different regions have different energy consumption patterns for cooking corresponding to different local conditions. The main types of fuels used by households from the mountainous areas (Aba) are firewood and electricity, as they have access to nearby forest and hydropower resources. Biomass energy (consist of crops straw, firewood and biogas) is commonly used by the households located in hilly areas (Yibin). The possible reason for this could be that the weather and geographic conditions in that area are quite suitable for planting, forestry, and animal breeding. Thus, households have abundant biomass resources collected from residues of these productive activities. Finally in plain areas (Deyang), the use of commercial energy such as electricity, LPG, and natural gas is quite popular because of the relatively high development level of the regional economy.

In Figure 3.4, we divided our sampled households into two groups. One includes the traditional biomass energy users (who are using crops and firewood), and the other only consists of non-traditional biomass energy users (users of other energy). Figure 3.4 illustrates the comparison of the income levels between these two groups. It can be seen that the average income level of the non-traditional biomass energy users is higher than that of the traditional biomass energy users. This proves that income level is a rather important determinant for household cooking fuel choice. Higher income levels will reduce the household use of traditional biomass energy.

Figure 3.4 Income levels of two household groups

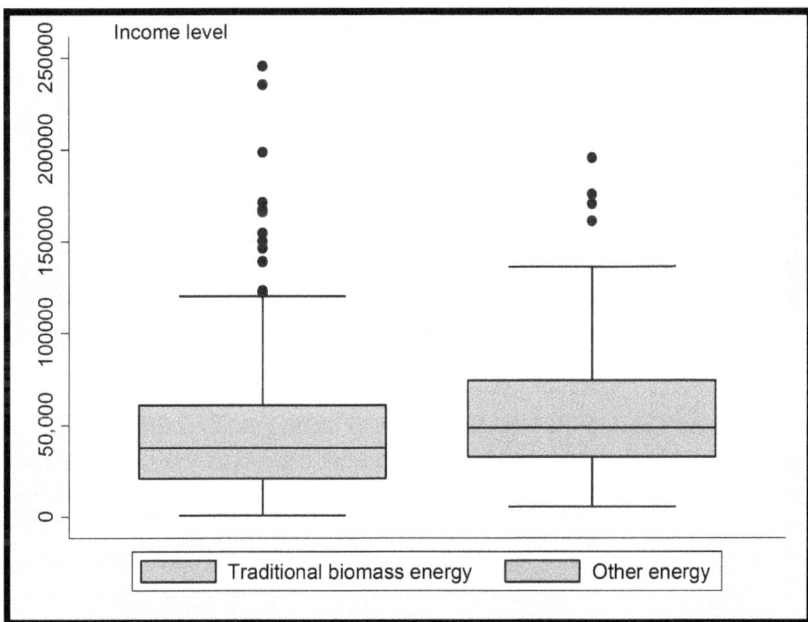

Source: Author's own field survey

In practice, traditonal biomass energy (consisting of crop straw and firewood), biogas, coal, and electricity are the main types of energy adopted by the surveyed households for cooking. Figure 3.5 shows the cooking stoves of the four main types of energy sources employed by the households in the study region. The cost of coal stove is the lowest, whereas the price of electric cookers is the highest. Cooking with electricity is the most efficient, while the time of cooking is the longest for traditional biomass energy. Among gaseous fuels, biogas has the lowest use cost.

Figure 3.5 Cooking stoves four main types of energy sources in study region

Note: From top left to bottom right: improved stove for traditional biomass energy (crop straw and firewood), biogas stove and cooker, coal stove and electric cooker.

Moreover, considering the household choices among the four different energy alternatives (namely traditional biomass energy, biogas, coal and electricity), the decision-making behaviors are household specific. Household characteristics, therefore, could impact the household fuel choices on cooking fuel use. Table 3.1 gives the characteristics of households selecting four different energy alternatives. The households who choose traditional biomass energy have the lowest fraction of adult female members (0.394), while having the highest fractions of children and elderly people (0.122 and 0.146). They also have the smallest mean value of per capita income (12355 CNY per year) and the largest area of arable land (4.672 Mu). On the contrary, households choosing electricity have the smallest size (3.91) and the highest average per capita income level (17289 CNY per year). They also have the highest fraction of female adult members (0.439) and the lowest fraction of children (0.094). Additionally, households picking biogas have the largest size (4.29) and the highest fraction of female adult members (0.443), whereas the households who select coal have the smallest size (3.960) and the lowest fraction of elderly people (0.089). Particularly, they own the smallest area of arable land (2.996 Mu). With respect to the accessibility of biomass resource, households selecting traditional biomass energy live the farthest from the nearest biomass collection spots (3.9 Km), while those choosing biogas are the nearest (0.5 Km).

Table 3.1 Comparison of characteristics of households choosing four different energy alternatives

	Traditional biomass energy	Biogas	Coal	Electricity
Family Size (Number)	4.22	4.29	3.96	3.91
Fraction of adult male members	0.439	0.443	0.407	0.437
Fraction of adult female members	0.394	0.416	0.415	0.439
Fraction of children (<=14)	0.122	0.115	0.120	0.094
Fraction of elderly (>=65)	0.146	0.095	0.089	0.114
Income per capita (CNY/Year)	12355	13512	14373	17389
Distance to the nearest biomass collecting spot (Km)	3.9	0.5	1.2	1.9
Arable land owned (Mu)	4.672	4.468	2.996	3.142

Source: Author's own field survey

Furthermore, energy choices could also be affected by the characteristics of decision makers (as it is assumed before, household heads usually make decisions on household cooking fuel choices). As is shown in Figure 3.6, traditional biomass energy and coal represent the larger share among female household heads than of male household heads; while a smaller share of female household heads prefer electricity and biogas.

Figure 3.6 Energy choices made by household heads with different gender

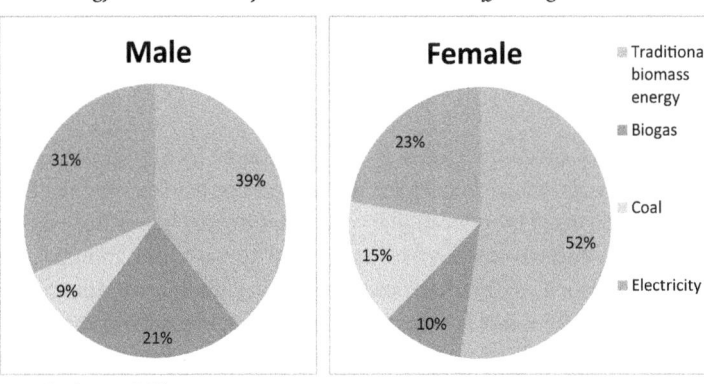

Source: Author's own field survey

The energy choices made by household heads from different age groups are illustrated in Figure 3.7. It can be seen that the older household heads are more likely to choose biomass energy (biogas and traditional biomass energy), whereas they have less likelihood to select coal and electricity for cooking. The proportion of household heads choosing traditional biomass energy is the largest in the oldest age group of 65+, while it is the smallest in the youngest age group of 15–34. The share of the household heads selecting electricity is the smallest in the age group of 65+, while it is the largest in the age group of 35–44. In addition, the proportion of the household heads picking coal is the largest in the age group of 15–34, whereas that of household heads choosing biogas is the largest in the age group of 55–64.

Figure 3.7 Energy choices made by household heads of different age groups

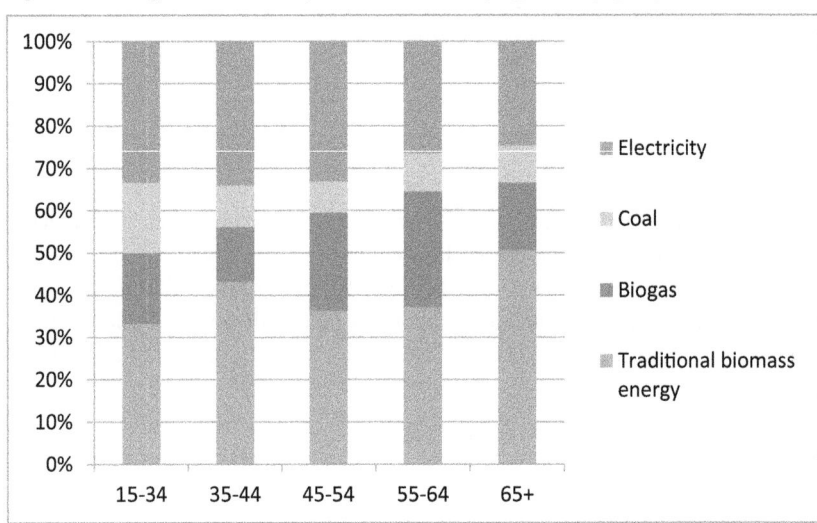

Source: Author's own field survey

Finally, in order to check the effects of the educational level of household heads on household energy choice behaviors, all sampled households were divided into three groups in terms of the educational years of household heads (See Figure 3.8). Obviously, well-educated household heads prefer to use clean and effective energy alternatives, such as electricity and biogas, over traditional biomass energy. The majority of the households who are illiterate or literate below primary school (educational level of 0–5 years) select traditional biomass energy, whereas most of the households with an educational level of high school and above (9 years+)

select electricity as cooking fuel. According to that described above, household head characteristics such as gender, age, and educational level are also important determinants of household cooking energy choices.

Figure 3.8 Energy choices made by household heads with different educational levels

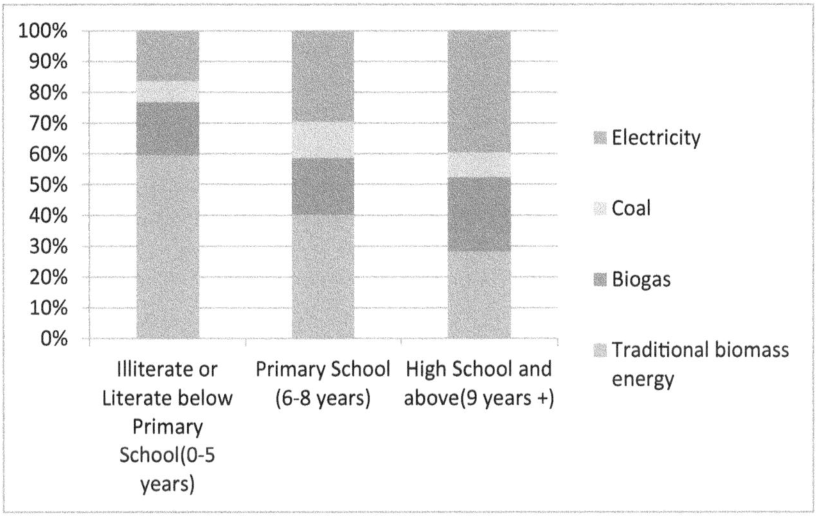

Source: Author's own field survey

3.4 Empirical Strategy

3.4.1 Revealed and Stated Preferences

Methodologies for examining the choice behaviors of consumers for certain products or services are numerous. However, in the existing literatures, these methodologies can be divided into two parts; i.e. revealed preference (RP) and stated preference (SP) approaches (Whitehead et al., 2007). They are often used to evaluate consumers' preferences, analyze their demands, and forecast their behaviors. The revealed preference approaches use observed actual choice data to estimate the ex-post willingness to pay (WTP) for different commodities, while the stated preference approaches examine consumers' preferences over a range of hypothetical alternatives based on SP choice experimental design to estimate the ex-ante willingness to pay for various commodities (Brownstone et al., 2000; Whitehead et al., 2007).

The RP methods are the most appropriate and effective tools for analyzing preferences, inferring utility functions, and estimating consumers' demands for the products or services based on a comparison of the chosen alternatives and the rejected alternatives (Kroes and Sheldon, 1988). The data used in RP analysis are usually obtain by direct observation of the consumption behaviors or collected in some surveys asking the respondents about their actual choices. However, the RP approaches have many weaknesses that restrict their general applications. Firstly, they rely heavily on historical data. Researchers are not allowed to model the preference or demand for new products or new government policies which lie beyond the range of the historical experience of RP data (Whitehead et al., 2007). Sometimes the RP data are also limited to analyze the real preferences of the individuals for some government-oriented projects. For instance, the RP data of biogas used in our research cannot reflect households' real preferences for biogas, because the construction of household-based biogas digesters in Sichuan Province is government driven. Secondly, there are often strong correlations between explanatory variables of interest and other variables which may lead to statistically insignificant coefficient estimates and imprecise estimates of attribute contributions to the utility functions (Kroes and Sheldon, 1988; Cameron et al., 1996; Brownstone et al., 2000; Mark and Swait 2004; Whitehead et al., 2007). In this situation, the multicollinearity and endogeneity problems usually occur in the econometric analysis. Thirdly, RP data are always obtained without sufficient variation among attributes in the real market. Therefore, it is difficult to use these data to examine all variables of interest (Kroes and Sheldon, 1988; Brownstone et al., 2000 Mark and Swait 2004). Fourthly, the RP data collection is sometimes inefficient, because the data can only be collected in the field survey at one time point, resulting in a relatively small sample size (Whitehead et al., 2007). In addition, the explanatory variables are also required to be expressed in objective units in RP methods. That is to say, RP data are normally restricted to some primary variables such as price, cost, and time (Kores and Scheldon, 1988). It is difficult to test some secondary variables (such as the impacts on environment and health) using only RP data.

Due to the limitations of RP data, SP methods have become an attractive option in preference research. They were originally developed in marketing research and have been widely used in the research field of transportation, marketing, environmental economics, and health economics. Researchers can analyze consumers' preferences for products or services that do not exist in the real market by using SP experiments to elicit their preferences over hypothetical alternatives (Louviere et al., 1983; 2000). The main advantages of SP methods are flexibility and the ability to measure non-use value (Whitehead et al., 2007). In SP experiments, the choice sets can be designed before implementing the experiments. The range of the attributes can be extended by setting hypotheses. Some qualitative attributes such as safety and smokiness level in energy studies can be included in the experimental design, and the multicollinearity among attributes could be avoided (Morikawa, 1989). Nevertheless, they still have some disadvantages, among which, one major problem is the hypothetical nature of SP data. The consumers behave differently in response to hypothetical attributes and their levels than they would when they face the same situation in the real market. In some cases, if the respondents are not well educated, they may not be able to completely understand the SP experiments (Brownstone el al., 2000). Another sever problem could arise when the new products have politically correct public good attributes such as zero-pollution energies. The respondents are highly likely to misunderstand these options in SP experiments and then to misrepresent their preferences for them, although in reality, they are unwilling to pay more for these types of energy (probably due to the free-rider problem) (Brownstone et al., 2000).

In order to simultaneously exploit the advantages of the data of each approach and mitigate the shortages, the RP methods and SP methods are usually combined to jointly estimate the consumers' preferences by many researchers, considering the complementary relationship between RP data and SP data. This new potential method was firstly adopted in the field of transportation and marketing (Morikawa 1989; Ben-Akiva and Morikawa, 1990; Morikawa et al., 1991; Hensher and Bradley, 1993; Swait et al., 1994; Hensher et al., 1999). And then it was applied in the research on environmental economics and health economics (Adamowicz et al., 1994; 1997; Huang et al., 1997; Mark and Swait 2004). The literatures relating to developing the joint estimation models for RP and SP data seek to enhance the contrasting strengths of both of these approaches while reducing their weaknesses. As mentioned in the work of Hensher et al. (1999) and Louviere et al. (2000), the RP and SP data combination can be regarded as a way of data enrichment. The RP data can be enriched by SP data. Combining RP data with SP data allows the models to extend beyond the limited range of historical experience and can be used to collect information from the current non-users of some products or services. When RP and SP data are collected based on the appropriate experimental design, the multicollinearity and endogeneity problems could be avoided by using these data (von Haefen and Phaneuf, 2007). Furthermore, the combination of RP data and SP data can obtain more information from each respondent to improve the data collection efficiency as well as econometric efficiency (Whitehead et al., 2007). The joint estimation of choice models can simultaneously take the hypothetical choices and the actual choices into consideration to mitigate the hypothetical bias caused by SP data. It can be also used to validate both types of the data (Whitehead et al., 2007). Thus, based on what has been stated in this section, this research will estimate the households' choice behaviors by using RP, SP, and joint RP-SP data, respectively, based on the random utility theory, which will be discussed in Section 3.4.3.

3.4.2 Model Specification

Following McFadden (1973; 1974), the basic utility theory in economics is employed to address the choice problem in this paper. Suppose that each household from the same population of interest faces a finite choice set and selects an alternative to maximize its utility.

We denote C_i as the choice set with J alternatives faced by household i and $u_i(\cdot)$ as its random utility function, assigning a value to each potentially available alternative. The household chooses some alternative $j \in C_i$ having a vector of measured attributes x_j. Therefore, the reveal preference of the household for a particular type of energy can be expressed as $u_i(x_j) \geq u_i(x_k)$ (all $k \in C_i$). Thus, the choice probability that j will be chosen can be derived as:

$$P(j \mid x_j) = P[i : u_i(x_j) \geq u_i(x_k), \text{ for all } k \neq j \in C_i] \tag{3.1}$$

Where the distribution of the utility function $u_i(x)$ is assumed to satisfy that the probability of ties is zero, and $x = (x_1, \ldots x_j)$. Equation (3.1) is also referred to as the explanation of observed choice in a random utility model (RUM). Its right-hand side represents the probability that a household chosen randomly from the population of interest selects an alternative j that maximizes its utility (Manski, 2001).

The random utility function $u_i(x_j)$ can also be interpreted as attainable maximum utility for the selected household, given its budget constraint and fixed alternative k, as outlined in McFadden (1980). Then $u_i(x_j)$ is a function of all alternatives in the choice set C_i. The randomness in the utility function mainly comes from the unobserved variations in tastes and attributes of different alternatives, and the errors of perception of households (McFadden, 1973; 1980).

Based on RUM, Equation (3.2) defines the random utility function of choice j, for household i, who are facing J unordered alternatives:

$$U_{ij} = X_{ij}\beta + (z_{ij}\gamma)' + \varepsilon_{ij} \tag{3.2}$$

Where X_{ij} is a matrix of alternative-specific variables, z_{ij} is a matrix of case-specific variables, ε_{ij} is the disturbance term, and β and γ are the matrices of regression coefficients.

If the household makes a particular choice j, then we assume that U_{ij} is the maximum utility among all the existing J choices. Hence, the probability that choice j is made can be expressed as:

$$\text{Prob}(U_{ij} > U_{ik}), \quad k \neq j \tag{3.3}$$

Let Y_{ij} be a random variable that indicates the choice made. That is, $Y_{ij} = 1$ if the household i chooses the alternative j and $Y_{ij} = 0$ otherwise. According to McFadden (1973), if (and only if) the J disturbances are independent and identically distributed with Gumbel (type 1 extreme value) distribution, namely:

$$F(\varepsilon_{ij}) = \exp(-\exp(-\varepsilon_{ij})) \tag{3.4}$$

Then

$$\Pr ob(Y_{ij} \mid X_{ij}, z_{ij}) = \frac{\exp(X'_{ij}\beta)\exp(z'_{ij}\gamma)}{[\sum_{j=1}^{J}\exp(X'_{ij}\beta)]\exp(z'_{ij}\gamma)} \tag{3.5}$$

This leads to what is called the Alternative-specific Conditional Logit model (asclogit) or McFadden's choice model (McFadden, 1973; Greene, 2012). This paper uses this model to test the household energy choice behaviors in rural Sichuan.

3.5 Empirical Analysis

3.5.1 Revealed Preference (RP) Analysis

In this research, the households' revealed preferences for energy choices are discussed firstly. The concept of revealed preference (RP) was introduced by Samuelson (1938). It asserts that the best way to measure consumers' preferences is to observe their purchasing behavior on the basis of the assumption, given that a consumer chooses one option out of a set of different alternatives, this option must be the preferred option (Samuelson, 1938). Then, in an attempt to model the decision process of the household, the choice modeling methods are applied to analyze the RP data (McFadden, 1973).

3.5.1.1 Household Energy Choice and Its Determinants

We have collected the data on 556 households and their choices among 4 different types of energy for cooking by asking respondents questions like "Currently, which type of energy do you prefer the most for cooking?" The energy types are traditional solid biomass energy consisting of crop straw and firewood, biogas, coal, and electricity.

The actual choices made by the surveyed households are presented in Figure 3.9. Considering lower-income households, an obvious preference for traditional solid biomass energy over all the remaining three types of energy is shown, as 129 (69.7%) of these households choose firewood for cooking. 13.51% uses coal and 14.05% uses electricity. Only 5 households use biogas as their main cooking energy, accounting for a rather small share (2.7%). The main reason for this situation could be that these households are located in the mountainous areas. The local weather and geographical conditions of mountainous areas are not suitable for the construction and operation of biogas digesters. Concerning middle-income rural households, biomass energy plays a vital role in their cooking energy utilization. The majority of the households (70.4%) decide to choose biomass energy as their cooking fuels. Among which, 35.5% (66) households are still willing to use traditional solid biomass energy, whilst 32.8% (61) households select biogas. This is probably because these households live in hilly areas and depend on traditional agricultural production, bamboo planting, and the pig breeding industry for a living. Thus, they have abundant biomass resources such as crop straw, firewood, and animal dung for energy use. There are merely 18 (9.68%) households who choose coal to cook their food and boil water, whereas 41 (22.04%) households made their choice to consume electricity. This could be attributed to the relatively high level of electricity and coal prices faced by the households. For the high-income zone, electricity is preferred for cooking by a total of 103 (55.68%) households. On the contrary, 27 (14.59%) households chose traditional solid biomass energy and 48 (25.95%) chose biogas. The potentially crucial factor that is responsible for this phenomenon could be household location. Living in plain areas increases opportunities to get jobs with higher income in nearby towns or surrounding cities to afford commercial energy. In turn, households are unwilling to spend extensive time on biomass collection, which constrains them from participating in alternative income-generating activities. Compared to the households from low- and middle-income zones, the upper-income households are more likely to choose the type of energy on the higher level of the energy ladder. Therefore, in Sichuan Province, household location and regional economic development level could be the important determinants of household energy choice behaviors.

Figure 3.9 Household energy choices (Unit: households)

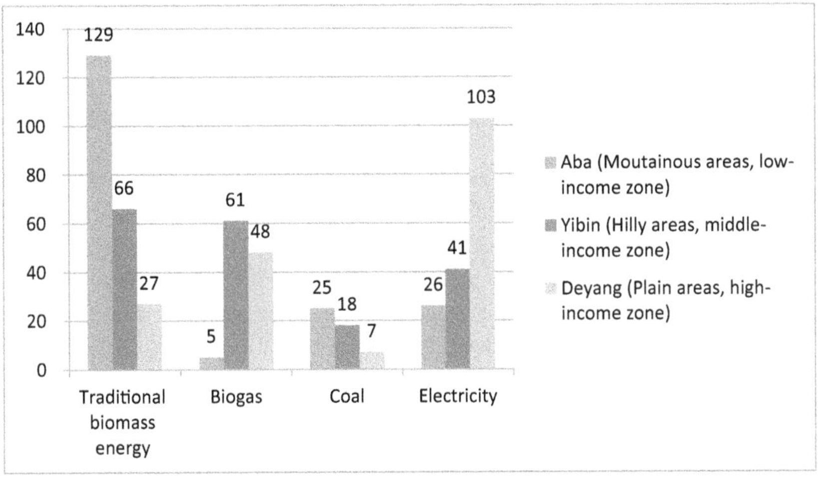

[Bar chart showing household energy choices across three regions:
- Traditional biomass energy: Aba 129, Yibin 66, Deyang 27
- Biogas: Aba 5, Yibin 61, Deyang 48
- Coal: Aba 25, Yibin 18, Deyang 7
- Electricity: Aba 26, Yibin 41, Deyang 103

Legend:
- Aba (Moutainous areas, low-income zone)
- Yibin (Hilly areas, middle-income zone)
- Deyang (Plain areas, high-income zone)]

Source: Author's own field survey

3.5.1.2 Descriptive Analysis of Variables Used in RP data Analysis

The explanatory variables used in analyzing the revealed preferences of the households are listed in Table 3.1. Energy price (EP) is selected as the alternative-specific (energy-specific) variable of the choice model.[9] Theoretically, the demand of one type of energy decreases as its price increases. As the biomass (such as crop straw, firewood, and animal dung) is non-tradable in Sichuan Province, the opportunity costs or shadow prices determined by the time spent on biomass collection and pigsty cleaning are taken as proxies for the prices of traditional biomass energy and biogas. Assuming that the rural labor market is perfect, we follow Kanagawa and Nakata (2007) to calculate the shadow prices of the biomass energy using the formula in Equation (6):

$$P_{ij} = \frac{w_i}{C_{ij}} \times T_{ij} \quad j = A, B \tag{3.6}$$

---

9 We also selected the smokiness level and safety risk as alternative-specific variables and treated them as pseudo-categorical variables (similar to process discussed in Section 3.3.2.1 of SP experimental design). However, due to the problem of multicollinearity, they were eliminated by Stata.

Where P_{ij} (CNY/kgsce[10]) is the shadow price of biomass energy j for household i; w_i (CNY per hour) is the market wage rate of household labor; C_{ij} (kgsce per year) is the household consumption of energy; and T_{ij} (Hours) is the time spent on biomass collection or pigsty cleaning. Particularly, the missing values for energy prices in the sample are replaced with the regional mean energy price values.

With respect to the case-specific (household-specific) variables, we assume that the household head is the decision-maker of actual cooking fuel choices. Therefore, characteristics such as the age, gender, and educational level of the household head (AHH, GHH, and EHH) will be regarded as important factors affecting household energy choice behaviors in RP analysis. Particularly, it is conjectured that the higher the decision-maker's education level is, the lower the probability of using traditional biomass energy will be (Ouedraogo, 2006; Farsi, et al., 2007; Mekonnen and Köhlin, 2008; Vijay and Aditi, 2011).

Among other house-specific variables, household family size (FS) is directly related to household energy choice behaviors through actual demand and available labor for biomass collection (Wambua, 2011). Households of larger size could have more labor resources. This also indicates that they need to make decisions on energy use to cook more food. Considering the demographic characteristics (such as the fractions of adult male (FAM), adult female (FAF), children (FC), and elderly people (FE)) are used to measure the amount of available labor resources that can be provided by a household. They also reflect the effects of the gender and age structure on household fuel choices to some extent. For a household, the higher the fractions of adult males and females are, the more likely that household is to use commercial energy (such as coal and electricity), whereas the higher the fractions of children and elderly people are, the greater the likelihood the household will use traditional biomass energy for cooking.

In order to capture the effects of income level on household cooking fuel switching decisions, the log-transformed value of income per capita (IPC) is used as a regressor in our models. As the income from off-farm work takes the largest share in total household income, the average income level could be regarded as the exogenous variable in our model. It is expected to have a positive impact on choosing high-quality energy. The areas of arable land owned by a household (AL) could matter in deciding the type of energy to be adopted for

---

10 In order to simplify our analysis, we unify different energy units to standard coal equivalent unit (kgsce) through dividing the energy consumption amount by the conversion coefficients.

cooking, as the crop residues from the cultivated land would be the major biomass resources for energy use. It is hypothesized that as the areas of arable land owned increases, the probability of usage of traditional biomass energy increases accordingly.

The variable used to reflect the impact of cooking habits on household energy choice behaviors is the "number of people frequently eating at home" (NPEH). In this study, we define the people frequently eating at home as household members who eat at home more than 5 days a week. This variable is expected to have a positive impact on the choice of clean fuels (coal and electricity).

Meanwhile, the distance from the house to the nearest biomass collecting spot (DB) is selected to represent the accessibility and availability of biomass resources. It is supposed to negatively influence household labor allocation to biomass collection, and thus, in turn, negatively impact the choice of traditional biomass energy. For the missing of this variable caused by non-participation in biomass collection, we assume that these households face the average distance and substitute the regional sample mean for the missing data.

In addition, household location could play an important role in household decision-making towards cooking fuel choices. The local dummy variables (r1 and r2) are constructed to test the effects of regions. They take a value of 1 if the surveyed households are from mountainous areas or plain areas, and 0 otherwise. It is conjectured that households located in mountainous areas are more likely to choose traditional biomass energy, whereas those from plain areas are more likely to use the fuels on higher levels of the energy ladder.

Table 3.2 Description of explanatory variables used in asclogit model for RP data

Variables	Mean	Std. Dev.
Alternative-specific variables		
Price of traditional biomass energy (CNY/kgsce)	1.2689	1.9259
Price of biogas (CNY/kgsce)	0.5631	0.4246
Price of coal (CNY/kgsce)	0.7758	0.0046
Price of electricity (CNY/kgsce)	4.0587	0.6656
Case-specific variables		
Gender of household head (Male=1, Female=0)	0.9281	0.2586
Age of household head (Years)	51.6727	11.6792
Educational years of household head (Years)	6.4245	3.4781
Income per capita (CNY/year)	14312.58	13754.38
Family size (Number)	4.1151	1.3667
Fraction of adult female	0.4146	0.1490

Variables	Mean	Std. Dev.
Fraction of adult male	0.4385	0.1645
Fraction of children (≤14)	0.1119	0.1560
Fraction of elderly member (≥65)	0.1207	0.2336
Land areas owned (Mu)	4.0117	3.7063
Distance to the nearest biomass collecting spot (Kg)	2.3349	4.3866
Number of people frequently eating at home (Number)	3.0935	1.3792
=1, if the household is from mountainous areas	0.3327	0.4713
=1, if the household is from plain areas	0.3327	0.4713
	556	

Note: CNY is abbreviation of Chinese Yuan. The missing dummy for region is "hilly area."

Table 3.2 reveals that electricity has the highest average price level at 4.06 CNY/kgsce, whereas biogas has the lowest average price level at 0.56 CNY/kgsce. The calculated average shadow price of traditional biomass energy for the household is 1.27 CNY/kgsce, which is approximately twice as much as the price of coal at 0.78 CNY/kgsce.

With respect to the case-specific variables, in our sample, the average family size of households is about 4.12. The average fraction of male and female adults is around 41.46% and 43.85%, while the average fractions of children and elderly people are 11.19% and 12.07%, respectively. The surveyed households are randomly selected from three different regions. The number of households from the mountainous and plain areas both amount to 33.27%, and 33.46% of the households are from hilly areas. The average income per capita of our sampled households is 14312.58 CNY per year. Moreover, according to the information listed in Table 3.2, 92.8% of the sampled households have male heads. The average age of the household heads is 51.67 years, and the average educational level is approximately 6.42 years. Furthermore, on average, the areas of arable land possessed by the households in our sample are 4.01 Mu. Specifically, within the 556 households, about 3.09 people frequently eat at home.

3.5.1.3 Results of Econometric Analysis

The estimation results of the asclogit analysis of the revealed preference data are presented in Table 3.3. Traditional biomass energy is the omitted category (the basic alternative), with which the estimated coefficients are to be compared (Mekonnen and Köhlin, 2008). The odds ratios are also included in the model to make interpretations easier. In order to better understand the substitution patterns between the four types of cooking fuels amongst the different households, the marginal effects of the statistically significant variables at sample means are calculated and presented in Table 3.4. The numbers in this table illustrate the effect of a one-unit change in a given independent variable (or a switch in the case of a dummy variable) on the probability of choosing a particular cooking energy source. In particular, the values of the predicted probability (See Table 3.4) confirm that most households have a high probability (43.7%) of choosing traditional biomass as cooking fuel. Households with a probability of adopting electricity account for 34.7% of the total sample, while those that could have a probability of selecting biogas and coal take up a small share (14.5% and 7.1%) of the population. This means that the sampled households still have a high propensity to choose traditional solid biomass energy for cooking. Meanwhile, electricity is preferred over biogas and coal.

Table 3.3 Estimation results of asclogit model for RP data

Alternative-specific variables	Coef.		Odds ratios			
Energy Prices (log)	−0.5147(0.08)***		0.5977			
Choice	Biogas		Coal		Electricity	
Case-specific variables	Coef.	Odds ratio	Coef.	Odds ratio	Coef.	Odds ratio
Gender of household head	0.6231(0.67)	1.8614	−0.3583(0.59)	0.6988	0.5420(0.57)	1.7195
Age of household head	−0.0076(0.01)	0.9924	0.0091(0.18)	1.0091	−0.0244(0.01)*	0.9759
Educational level of household head	0.0993(0.05)**	1.1044	0.0999(0.06)*	1.1051	0.1148(0.41)***	1.1216
Family size	0.1681(0.13)	1.1830	−0.0892(0.16)	0.9146	0.0059(0.12)	1.0060
Fraction of adult male	2.5277(1.70)	12.5245	0.1029(1.60)	1.7770	2.0997(1.43)	8.1637
Fraction of adult female	1.7536(1.72)	5.7754	1.0290(1.65)	4.6285	2.4091(1.45)*	11.1239
Fraction of children	1.2709(1.73)	3.5640	0.4863(1.65)	1.6262	1.6017(1.46)	4.9613
Fraction of elderly people	−2.9051(0.72)***	0.0547	−1.8713(0.91)**	0.1539	−1.8450(0.60)***	0.1580
Income per capita (log)	−0.0173(0.23)	0.9829	0.2259(0.28)	1.2534	0.4659(0.22)**	1.5934
Number of people eating at home	0.0206(0.13)	1.0208	0.0959(0.16)	1.1007	0.1061(0.12)	1.1119
Land areas owned	0.0268(0.04)	1.0271	−0.1987(0.76)***	0.8198	−0.1176(0.05)**	0.8891
Distance to the nearest biomass collecting spot	−0.3404(0.12)***	0.7115	−0.3837(0.13)***	0.6814	−0.0073(0.04)	0.9927
Located in mountainous areas	−2.6663(0.53)***	0.0370	−0.1708(0.43)	0.8430	−1.7982(0.37)***	0.1656
Located in plain areas	1.3961(0.38)***	1.5344	−0.1936(0.56)	0.8240	1.8737(0.36)***	6.5122
_cons	−3.1287(2.80)	0.1227	−3.0714(3.17)	0.0464	−5.3967(2.58)**	0.0045
Observations	556		556		556	
Wald Chi2(43)	221.98***					
Log likelihood	−527.38944					

Note: The missing dummy for region is "hilly area." The basic alternative in this asclogit model is "traditional biomass energy." Significant level: *10%, **5%, ***1%. The '(log)' means that variables are in logarithms.

Table 3.4 Marginal effects of key influencing factors in RP asclogit model

Alternative-specific variables	Traditional biomass	Biogas	Coal	Electricity
	Pr=0.437	Pr=0.145	Pr=0.071	Pr=0.347
Price (log)				
Traditional biomass	−0.1267***			
Biogas	0.0327***	−0.0639***		
Coal	0.0160***	0.0053***	−0.0340***	
Electricity	0.0780***	0.0259***	0.0127***	−0.1166***
Case-specific variables				
Age of household head	0.0039	0.0002	0.0013	−0.0054*
Educational level of household head	−0.0268***	0.0056	0.0027	0.0185**
Land areas owned	0.0223**	0.0113**	−0.0105**	−0.0231**
Distance to the nearest biomass collecting spot	0.0346***	−0.0379***	−0.0216***	0.0249**
Mountainous areas[a]	0.4021***	−0.1968***	0.0504*	−0.2558***
Plain areas[a]	−0.3430***	0.0619*	−0.0659**	0.3471***
Income per capita (log)	−0.8765*	−0.0279	0.0036	0.1008**
Fraction of elderly people	0.5221***	−0.2483***	−0.0482	−0.2256*

Notes: The significant level: *10%, **5%, ***1%. For dummy variables (a), the effects are obtained from probability differences. Pr is the predicted probability that each type of cooking fuel is chosen by a household.

The regression results reveal that energy price has a significant impact on household energy choice. The negative sign of its coefficient indicates that households prefer fuels with lower prices. In other words, raising the price of a type of energy decreases the likelihood that a household will use this type of energy. The marginal effects for energy prices provide more information about their effects. These results demonstrate that higher own prices of fuels are associated with significant negative shifts away from these fuels. In addition, the significant cross-price effects also reveal the substitutability between the different energy sources. According to the calculated marginal effects, a 10% increase in electricity price will decrease the average share of electricity users by about 1.17%, while increasing the average shares of traditional biomass energy, biogas, and coal users by about 0.78%, 0.26%, and 0.12%, respectively. A 10% decrease in coal price will increase the average share of coal users by about 0.34%, while decreasing the average share of traditional biomass energy and biogas users by around 0.16% and 0.05%, respectively. A 10% increase in the shadow price of biogas will decrease the average share of biogas users by about 0.64%, while increasing the traditional biomass energy share by approximately 0.33%. Finally, a 10% increase in the shadow price of traditional biomass energy will decrease the average share of its users by about 1.27%. The above results are consistent with our expectation.

Turning to the case-specific variables, income level, age and education level of household head, areas of owned arable land, household location, distance to the nearest biomass collecting spot, and fraction of elderly people in household members are found to significantly influence household choices on some, but not all, energy categories. Concretely, raising the household income level increases the probability of choosing commercial energy (coal and electricity) and simultaneously decreases the probability of adopting biomass energy (traditional biomass energy and biogas). For instance, a 10% increase in income per capita will reduce the shares of traditional biomass energy and biogas users by about 8.77% and 0.28%, respectively, while increasing the shares of coal and electricity users by about 0.04% and 1.01%, respectively. Regarding household head characteristics, the educational level has significant effects on household energy choices. The well-educated household head increases the probability of using fuels such as coal, biogas, and electricity, whereas those households in which the head has a lower educational level are more likely to choose traditional biomass energy. It suggests that increasing the educational level of the household head increases a household's likelihood to consume the types of energy of higher quality. Moreover, an increase in the age of the household head will decrease the probability of applying modern fuel (electricity) and increase the probability of choosing other energy sources. Among other household characteristics, the fraction of elderly people in household members plays an important role in determining energy choice behaviors. The signs of the marginal effects for it indicate that households with a larger fraction of elderly people prefer to use traditional biomass energy. This is because elderly people usually have more spare time to collect biomass and therefore face a lower opportunity cost for time. Households living farther from the place where they collect biomass are more likely to select traditional biomass energy and electricity as cooking fuels, whilst they are less likely to employ biogas and coal. Moreover, households who own more arable land are more likely to participate in agricultural production, thus they have more biomass resources collected from the agricultural residues. As a result, their probabilities of using biomass energy will increase, meanwhile that of purchasing commercial energy will decrease. Additionally, the geographical location is another important factor that affects household energy choice behaviors. This also suggests that there are differences in the energy choice behaviors of households in different areas. Compared to those from hilly areas, the households who live in mountainous areas are about 40.2% more likely to use traditional biomass energy and about 19.7% less likely to use biogas. There are several plausible reasons for this in the context of Sichuan Province: the relatively high price level of commercial energy in remote areas, the relatively education level of rural households for getting a good-salary job outside the village, the local weather conditions that are unsuitable for biogas production, and so on. On the contrary, the households located in plain areas are on average about 34.7% more likely than those located in hilly areas to use electricity and about 34.3% less likely to use traditional biomass energy, probably because of the higher awareness of environmental protection among local households and the higher development level of the local economy.

3.5.2 Stated Preference (SP) Analysis

In order to further investigate the influencing factors responsible for the biomass energy choice behaviors of rural households using the SP data, a labeled discrete choice experiment was designed. A discrete choice experiment is an increasingly popular methodology for the stated preference (SP) study. It helps to identify how individuals value particular attributes of a commodity, product, program, job, or policy by asking them to state their preference over hypothetical alternatives (Rao et al., 2012). The DCE method has some advantages over traditional study methods, which focus on the revealed preference (RP) data. It firstly provides a quantitative estimate of how a household values different attributes. Secondly, it allows comparison of several attributes with each other simultaneously, and it is fairly straightforward for households, as the choices closely resemble real-world decisions. In doing a DCE, a respondent is asked to choose an option from a set of alternatives under a specific situation. Recently, this method has been applied to study the energy choice behaviors in many different countries (Vaage, 2000; Braun, 2010; Shen and Saijo, 2009; Willis et al., 2011; Takama et al., 2012).

3.5.2.1 Experimental Design

- **Generation and refinement of choice options**

The first step in choice experiment design is to select and set up the choice options. The universal choice set was generated from China New Energy and Renewable Energy Statistic Yearbook (CRES, 2009) and China Energy Statistical Yearbook (CNBS, 2012), which show that 8 different types of energy (crop straw, firewood, biogas, coal, LPG, natural gas, solar, and electricity) are mainly used by rural households for cooking. However, this choice set with 8 energy choices is still too large to design a choice experiment. Therefore, it should be refined by removing the atypical and irrelevant options. In this study, there are three criteria for evaluating the relevance of the choice options: (1) the current rural energy consumption structure in Sichuan Province, i.e. the amount of each type of energy consumed by rural households; (2) the information collected from pilot survey and informal interview with the local energy experts from BIOMA (Biomass Institute of Ministry of Agriculture); and (3) the relevance to our research objectives. According to the statistics (CRES, 2009 and CNBS, 2012), the most commonly used types of biomass energy for cooking in rural Sichuan Province are crop straw, firewood, and biogas, while the major commercial energy sources used in villages are coal and electricity. Then, a total of 5 types of energy were selected as the significant existing options. From a pilot survey conducted in October 2013 and an informal interview with the energy experts, it is suggested that households always burn traditional biomass energy, such as firewood and crop straw, using the same improved brick stove. Thus, the number of choice options can be reduced by combining firewood and crop straw into one choice option of traditional biomass energy. Therefore, the four selected energy options for this cooking energy choice experiment are traditional biomass, biogas, coal, and electricity.

- **Identification of attributes**

Normally, the most important step for conducting a DCE is to identify the attributes and their levels. Hensher et al. (2005) provided three criteria for choosing the attributes; the attributes should (1) have significant impacts on the choice behaviors, (2) be invariant across all options, and (3) be related to the research objectives. In past works of research, a large number of attributes were reported to influence the household decision making on energy choice (Takama et al., 2012; Spautz et al., 2006). However, in order to make the experiment feasible and to reduce the complexity of the experiment, the number of the attributes has to be restricted (Lagarde and Blaauw, 2009). Thus, in line with the above three criteria and on the basis of a literature review, energy usage costs, device usage costs, degree of environmental pollution, smokiness level, and safety risk were initially chosen for the experiment. Based on the results of the household pilot survey, the attribute of degree of environmental pollution was excluded from the group of attributes, as it is not the main factor affecting the households' choices. Finally, four attributes were selected for the choice experiment. The pilot survey also revealed that each type of energy option in this experiment can just correspond to one type of stove in the real situation. And hence, the energy types were assigned as the labels of the choice options.

- **Assignment of attribute levels and labels**

Levels can be defined as the scale of measurement of a given attribute. The general information about the allocated levels and labels for the attributes of selected energy options is listed in Table 3.5. Two levels were assigned to one attribute in each energy option. Regarding to the two attributes of device and energy usage cost, the two levels correspond to the minimum and maximum cost for each energy option. To be more specific, the device usage cost (CNY/year) is calculated by allocating the cost of the stove over its life span. The stove cost was represented by its price, which was collected in the pilot survey. Herein, due to the data availability, the maintenance cost is not included in this indicator. For the commercial energy (coal and electricity), the energy usage cost refers to the product of energy consumption amount and energy price. In Sichuan Province, honeycomb coal briquette is the most common type of coal in rural areas. Therefore, the price of honeycomb coal briquette was used to represent the price of coal in this experiment, whilst the standard sales price of Sichuan Power Grid was considered to be the price of the electricity in this experiment. On the other hand, to measure the usage cost of biomass energy (traditional biomass energy and biogas), the shadow prices can be substituted for their opportunity costs (Mekonnen, 1999; Baland et al., 2010; Démurger and Fournier, 2011). Thus, the usage cost of biomass energy was expressed as the product of biomass collection time (hours per kg) and the net income of the rural household (CNY per hour). In addition, the energy usage cost also encompasses the level of stove/energy efficiency. It was comprehensively counted based on the data analysis of the pilot survey and the official statistics.

The level and label allocation processes for smokiness level and safety risk were more complicated due to the definition and measurement problems. Herein, the term "smoke" refers to the indoor smoke emitted by the stove. It includes not only the small particles in the smoke fog, but also the hazardous substances such as carbon oxides, sulfur oxides, and nitrogen oxides. The attribute of smokiness level can be treated as a pseudo-categorical variable (Takama et al., 2012). This is to say, almost no smoke=0, very little smoke=1, little smoke=2, relatively heavy smoke=3, heavy smoke=4. Analogously, for the safety risk attribute, the definitions of the levels are: (1) safe: almost no risk of burn, explosion, and poisoning; (2) little unsafe: almost no risk of explosion and poisoning but risk of burn; (3) moderately unsafe: risk of serious safety accidents; (4) unsafe: high risk of explosion, burn, and poisoning or frequent serious safety accidents. Thus, it can also be treated as several pseudo-categorical variables, i.e. safe=0, little unsafe=1, moderately unsafe=2, unsafe=3.

Table 3.5 Assignment of levels and labels for attributes in cooking energy choice experiments

	Traditional biomass(A)	Biogas(B)	Coal(C)	Electricity(D)
Device usage cost (CNY/year)	{30, 50}	{150, 220}	{6, 16}	{60, 80}
Energy usage cost (CNY/year)	{233, 389}	{73, 146}	{756, 1080}	{372, 564}
smokiness level	{relatively heavy smoke, heavy smoke}	{very little smoke, little smoke}	{little smoke, relatively heavy smoke}	{almost no smoke, very little smoke}
Safety risk	{moderately unsafe, little unsafe}	{unsafe, moderately unsafe}	{little unsafe, safe}	{moderately unsafe, unsafe}

Note: CNY is abbreviation of Chinese Yuan; the average exchange rate in November 2013 was 6.09 CNY/USD.

- **Construction of choice sets**

Having determined the relevant attributes and their levels, hypothetical energy alternatives with different combinations of attributes and levels could be constructed and presented to the individual respondents. In this research, a labeled design approach was used, though such designs are popular in the transport field (Hensher et al., 2005; Hensher, 2008). The names of the energy alternatives play the role of 'labels'. For this labeled design, the combination of attributes and levels resulted in an experimental design with 2^{16} different types of energy choice profiles. To simplify the sets shown to the respondents in the questionnaire, a common practice of using fractional factor design was taken to reduce the number of choice sets to a manageable size by ignoring the interaction effect (Hensher et al., 2005). Then, an orthogonal main effect design was applied with the help of software SPSS19.0. This procedure reduced the number of profiles to a level of 20 alternatives. However, this number was still considered too large for a respondent to handle (Louviere et al., 2000). To further reduce the cognitive burden of the respondent, all alternatives would be randomly divided into 5 groups with 4 alternatives in each group. Hence, the 'random number table' method was adopted. This method is much more effective than manually selecting the random samples. Firstly, the 20 alternatives were numbered consecutively from 1 to 20. Meanwhile, a random number table was generated by Excel using the command '=INT (20*RAND () +1)'. In this table, the numbers were randomly selected from within the range of 1 to 20. Starting with an arbitrary number as appeared in a certain row as the first number, the second number can be selected by proceeding across the row to the very next different number. Continuing along the row, additional numbers were selected. Particularly, duplicated numbers were not allowed. Finally, 4 numbers were allocated in one group. Thus, 20 random numbers were split into 5 groups. By matching the number of each alternative to each random number, all of the 20 alternatives, therefore, were distributed into 5 groups. At this time, the number of choice sets was reduced to 4. That means each respondent must face 4 choice sets and in each set he or she has to make choice among 4 energy alternatives. An illustration of a choice set is presented in Table 3.6 below:

Table 3.6 A sample of a choice set in the choice experiment

	A	B	C	D
Device usage cost per year (CNY)	50	150	6	80
Energy usage cost per year (CNY)	389	73	756	564
smokiness level	relatively heavy smoke	little smoke	little smoke	almost no smoke
safety risk	little unsafe	moderately unsafe	little unsafe	unsafe
Your choice	☐	☐	☐	☐
If you don't want choose any one from above set, please tick here. ☐				

Source: Author's own field survey.

3.5.2.2 Descriptive analysis of the variables used in SP data analysis

The explanatory variables that will be used in our econometric analysis are listed in Table 3.7. In the choice experiment, the decision-maker should be the respondents[11] to whom the choice sets were shown. Of all the respondents, 66.37% of them are male, and 92.81% of them are married. The respondents have an average age of 51.55 years old, with the average educational level of 6.07 years. As each respondent has to face four choice sets, we have 2224 observations in total. The other case-specific variables used in SP data analysis are the same as those adopted in the RP data analysis (See Table 3.2).

---

11   66.7% of the respondents are household head.

Table 3.7 Description of explanatory variables used in asclogit model for SP data

Variables	Mean	Std. Dev.
Energy usage cost (CNY/Year)	-	-
Device usage cost (CNY/Year)	-	-
Smokiness level (almost no smoke=0, very little smoke=1, little smoke=2, relatively heavy smoke=3, heavy smoke=4)	-	-
Safety risk (safe=0, little unsafe=1, moderately unsafe=2, unsafe=3)	-	-
Gender of the respondent (Male=1, Female=0)	0.6637	0.4725
Age of the respondent (Years)	51.5522	12.4701
Educational years of the respondent (Years)	6.0719	3.6050
Marital status of the respondent (Married=1, Others=0)	0.9281	0.2584
No. of Obs		2224

Notes: The levels of the alternative-specific attributes are listed in Table 4, and the same case-specific variables that have been used in RP data analysis are listed in Table 2.

3.5.2.3 Results of econometric analysis

The alternative-specific conditional logit model was estimated using the SP data (presented in Table 3.8). Similarly, the marginal effects of statistically significant influencing factors at sample means are calculated and listed in Table 3.9. The predicted probability values (See Table 3.9) demonstrate that households having a probability of choosing coal account for the largest share (39.8%) of the population, suggesting that households have the highest potential preferences for coal over other types of energy sources. Households with a probability of adopting traditional biomass energy and biogas respectively make up 21.6% and 28% of total sample, while those that could have a probability of selecting electricity occupy the smallest share (10.6%) of the population.

According to the estimation results of the asclogit model, all the signs of the alternative-specific variables are consistent with our theoretical expectation. The statistically significant coefficients clearly show that device usage cost, smokiness level, and safety risk are the important determinants of household cooking fuel choices. The usage costs of the energy and the corresponding stove have negative impacts on household energy choice behavior, indicating that households prefer to use energy with lower usage costs. The negative coefficient of smokiness level reflects that households incline to choose the energy that could generate less smoke in the kitchen, whilst the positive coefficient of safety risk indicates that households have high preferences for safe energy carriers. More specifically, the marginal effects of the key influencing factors demonstrate that a 10% increase in the usage cost of electric cookers will decrease the average share of electricity users by about 0.16%, while increasing the average shares of traditional biomass energy, biogas, and coal users by about 0.04%, 0.05%, and 0.07%, respectively. A 10% decrease in the stove usage cost of coal will increase the average share of coal users by about 0.4%, while decreasing the average share of traditional biomass energy and biogas users by around 0.14% and 0.19% respectively. A 10% increase in the usage cost of biogas stove will decrease the average share of biogas users by about 0.34%, while increasing the traditional biomass energy share by approximately 0.1%. In addition, a 10% increase in the usage cost of the improved stove for traditional biomass energy will decrease the average share of its users by about 0.28%. Taking measures to reduce the smokiness level of a type of fuel increases the probability of choosing this type of fuel, while decreasing the probability of using other types of energy sources. On the contrary, improving the safety of a type of energy increases the chance for a household to use this type of energy, while lowering the chance to use other types of energy sources.

Considering the case-specific variables, income level is an important determinant. The marginal effect of income indicates that there is a lower chance for a household to use inferior types of energy sources (traditional biomass energy, biogas, and coal) whilst demonstrating a higher chance to choose electricity with an increase in income level. It also reveals that a 10% increase in income per capita will increase the average share of electricity users by about 0.48%, while reducing the average shares of traditional biomass energy, biogas, and coal users by about 0.06%, 0.18%, and 0.23%, respectively. The characteristics of the respondents, except gender, are factors affecting household energy choice behaviors. The educational level of the respondents significantly impacts household choices in all energy categories. The respondents with higher educational level are more likely to choose higher-quality energy, especially coal and electricity over the traditional biomass energy. Increasing the educational level of the respondent will increase the probability of choosing commercial energy (coal and electricity), while decreasing the probability of using biomass energy (traditional biomass energy and biogas). For older respondents, the other types of energy could provide higher utility to them than the traditional solid biomass. Particularly, with an increase in the age of the respondent, the probability of using biogas significantly increases.

In terms of the household characteristics, the significant coefficients of the asclogit model listed in Table 3.8 also indicate that demographic structure variables (fractions of children, male adults, and female adults) are main factors influencing household energy choice behaviors. Among which, the fraction of children can significantly affect the probability of choosing all types of energy sources. Higher fractions of children or adults imply that more labor would be available for biomass collection and energy preparation, thus simultaneously increasing the probability of using biomass energy and decreasing the probability of choosing commercial energy. Furthermore, the larger the number of people frequently eating at home, the lower the probability is that the household uses biogas. This is probably because households are unwilling to bear the higher costs of feeding more pigs for biogas production. On the other hand, due to the considerable transportation costs, the longer the distance to the nearest biomass collecting spot, the lower the probability that biomass energy is being used for cooking.

With respect to household location, the model regression results illustrate that households that live in mountainous areas are less likely to select biogas and electricity over traditional biomass energy than those living in hilly areas, but they are more likely to choose coal over traditional biomass energy than those from hilly areas. Households from plain areas are more likely to choose higher quality energy over traditional biomass energy than those from hilly areas. The marginal effects of location dummies also indicate that households that live in mountainous areas are respectively 12.2% and 6.1% less likely to select biogas and electricity than those living in hilly areas and about 0.5% and 17.8% more likely to choose traditional biomass energy and coal, respectively. Households from plain areas are approximately 29% more likely to choose coal than those from hilly areas and around 15.5%, 14.7%, and 1.2% less likely to use traditional biomass energy, biogas, and electricity, respectively.

Table 3.8 Estimation results of asclogit model for SP data

Alternative-specific variables (Energy)	Coef.	Odds ratios	Coef.	Odds ratios	Coef.	Odds ratios
Energy usage cost (log)	−0.1242(0.10)	0.8832			−0.1683(0.05)***	0.8451
Device usage cost (log)	−0.1679(0.07)**	0.8454			0.4269(0.05)***	1.5325
	Biogas		Coal		Electricity	
Case-specific variables (Choice)	Coef.	Odds ratio	Coef.	Odds ratio	Coef.	Odds ratio
Gender of respondent	−0.0982(0.14)	0.9065	0.1131(0.13)	1.1197	0.2344(0.19)	1.2642
Age of respondent	0.0130(0.01)*	1.0131	0.0072(0.01)	1.0072	0.0111(0.01)	1.0111
Educational level of respondent	0.0431(0.02)**	1.0441	0.1550(0.02)***	1.1677	0.1281(0.03)***	1.1367
Income per capita (log)	−0.0366(0.10)	0.9640	−0.0291(0.10)	0.9713	0.4793(0.13)***	1.6149
Land areas owned	0.0048(0.02)	1.0048	−0.0053(0.02)	0.9948	0.0121(0.02)	1.0122
Family size	−0.0867(0.06)	0.9169	−0.0317(0.06)	0.9688	0.0063(0.08)	1.0063
Fraction of elderly people	−0.0658(0.35)	0.9363	0.5530(0.34)	1.7384	0.2422(0.45)	1.2741
Fraction of children	−0.3733(0.76)	0.6884	−3.0380(0.70)***	0.0479	−3.7091(0.90)***	0.0245
Fraction of adult males	−1.1302(0.75)	0.3230	−2.0800(0.68)***	0.1249	−3.9407(0.87)***	0.0194
Fration of adult females	−0.4015(0.77)	0.6693	−2.0700(0.70)***	0.1262	−3.9005(0.89)***	0.0202
Number of people eating at home	−0.1222(0.05)**	0.8850	−0.0245(0.05)	0.9757	−0.0384(0.07)	0.9623
Distance to the nearest biomass collecting spot	0.0127(0.02)**	1.0127	0.0628(0.01)***	1.0648	0.0158(0.03)	1.0159
Mountainous areas	−0.4995(0.16)***	0.6069	0.3986(0.16)**	1.4897	−0.6585(0.23)***	0.5176
Plain areas	0.2452(0.18)	1.2779	1.5146(0.17)***	4.5478	0.9587(0.21)***	2.6082
_cons	1.4467(1.23)	4.2489	0.2646(1.17)	1.3030	−3.0245(1.56)*	0.0486
No. of Obs.	2221		2221		2221	
Wald Chi2(46)	410.53***					
Log likelihood	−2648.6128					

Note: The missing dummy for region is hilly area. The basic alternative for this asclogit model is traditional biomass energy. The significant level: *10%, **5%, ***1%. The '(log)' means that the variables are in logarithms. 3 cases (12 observations) were dropped by Stata due to no positive outcome per case.

Table 3.9 Marginal effects of key influencing factors in SP asclogit model

Alternative-specific variables	Traditional biomass Pr=0.2156	Biogas Pr=0.2797	Coal Pr=0.3984	Electricity Pr=0.1063
Device usage cost (log)				
Traditional biomass	−0.0284**			
Biogas	0.0101**	−0.0338**		
Coal	0.0144**	0.0187**	−0.0403**	
Electricity	0.0038**	0.0050**	0.0071**	−0.0160**
Smokiness level				
Traditional biomass	−0.0285***			
Biogas	0.0101***	−0.0339**		
Coal	0.0145***	0.0188***	−0.0403***	
Electricity	0.0039***	0.0050***	0.0071***	−0.0160***
Safty risk				
Traditional biomass	0.0722***			
Biogas	−0.0257***	0.0860***		
Coal	−0.0367***	−0.0476***	0.1023***	
Electricity	−0.0098***	−0.0127***	−0.0181***	0.0406***
Case-specific variables				
Educational level of respondent	−0.0189***	−0.0124***	0.0269***	0.0043*
Income per capita (log)	−0.0627	−0.0184	−0.0232	0.0478***
Fraction of children	0.3685***	0.3736***	−0.5295***	−0.2126***
Fraction of adult males	0.3371***	0.1212	−0.2057***	−0.2526***
Fraction of adult females	0.2914***	0.2657**	−0.2862	−0.2709***
Number of people eating at home	0.0104	−0.0207**	0.0093	0.0010
Distance to the nearest biomass collecting spot	−0.0065***	−0.0049*	0.0130***	−0.0015
Mountainous areas	0.0047	−0.1220***	0.1778***	−0.0606***
Plain areas	−0.1554***	−0.1468***	0.2904***	−0.0118

Notes: The significant level: *10%, **5%, ***1%. For dummy variables (a), the effects are obtained from probability differences. Pr is the predicted probability that each type of cooking fuel is chosen by a household.

3.5.3 Joint estimation of RP-SP data

In this section, two joint asclogit models were estimated using the pooled RP and SP data. The only difference between these two models is that we included the household characteristics in the second model. The results are listed in Table 3.10 (The second model is on the right side and the basic alternative for both of these models is traditional biomass energy). For energy-specific variables, we eliminated the device usage cost due to the unavailability of the data in RP analysis. However, the signs of the rest three variables are not only the same to those in SP model, but also in accordance with our expectations. The smokiness level and safety risk are still significant in the joint models. This is probably because the collinearity in the RP model has been reduced to some extent.

On the other hand, in terms of household-specific factors, it can be seen that the majority of the parameters have the same sign as the parameters in the SP model. Nevertheless, the fraction of children and the fraction of elderly people in the household have different signs from those in the RP and SP models. This may be caused by the differences between the structures of RP and SP data (Swait and Louviere, 1993; Louviere et al., 2000). According to previous studies (Adamowicz et al., 1994; Earnhart, 2001; Mark and Swait, 2004), a test procedure should be involved to check the structural changes. Firstly, the asclogit models are separately estimated using RP and SP data to get the likelihood ratios for (L_r and L_s) of these models (See Tables 3.4 and 3.8). Second, the two data sets are combined together to estimate a joint asclogit model to get the pooled likelihood ratio, which is denoted by L_p (See Table 3.10). Then, the null hypothesis of equal parameters was tested using the likelihood ratio test statistic: $\lambda = 2[L_p - (L_r + L_s)]$. Failure to reject this chi-square test means that the stated and revealed data have similar preference structures (Earnhart, 2001; Mark and Swait, 2004). In this study, the chi-square statistic for the test of equal parameters is about 362.580 (317.410 for the second model), which is much higher than the critical chi-square value, given 14 (or 16) degrees of freedom and the significant level of 1%, thus implying that the parameter equality hypothesis has to be rejected. In other words, the SP data and RP data contain different preference structures. The plausible reason for this is that the impact of the socio-demographic variables (e.g. the characteristics of the decision makers) is significantly different in the two data sources (Swait and Louviere, 1993; Louviere et al., 2000; Mark and Swait, 2004). Actually, in our research, the household head was assumed to be the decision maker in the RP model, whereas the respondent in the discrete choice experiment was treated as the decision maker in the SP model. Therefore, the RP data and SP data have different structures due to the characteristics of different decision makers. In this situation, combining the two data sources cannot thoroughly eradicate the underlying differences as described in Section 3.2 between them (Earnhart, 2001). That is to say, we cannot get a general conclusion on the impact of the factors such as fractions of elderly people and children on household energy choice from the joint models.

Table 3.10 Joint estimation results of the combination of RP and SP data

Alternative-specific variables	Coef.[a]	Coef.[b]					
Cost/Price (log)	−0.1218***	−0.1311***					
Smokiness level	−0.1578***	−0.1534***					
Safty risk	0.1079**	0.1037**					
Choice	Biogas		Coal		Electricity		
Case-specific variables	Coef.[a]	Coef.[b]	Coef.[a]	Coef.[b]	Coef.[a]	Coef.[b]	
Educational level of household head		0.0439**		0.1236***		0.0938***	
Age of household head		0.0079		−0.0001		−0.0040	
Gender of household head		0.1688		−0.1689		0.3072	
Income per capita (log)	−0.0071	−0.0413	0.1442*	0.0514	0.5904***	0.5253***	
Family size	−0.0220	−0.0318	−0.0293	−0.0381	−0.0193	−0.0208	
Fraction of elderly people	−0.4367	−0.5396*	−0.0473	0.0598	−0.3510	−0.2271	
Fraction of children	0.2321	0.2356	−2.2156***	−2.3571***	−1.9971***	−2.0994***	
Fraction of adult males	−0.2270	−0.2569	−1.4791***	−1.3652**	−2.2252***	−2.1926***	
Fraction of adult females	0.1304	0.1125	−1.5261***	−1.5012***	−1.9261***	−1.8474***	
Land areas owned	0.0083	0.0115	−0.0121	−0.0044	−0.0154	−0.0105	
Number of people eating at home	−0.0948**	−0.0950**	0.0121	−0.0004	0.0271	−0.0053	
Distance to the nearest biomass collecting spot	−0.0234	−0.0154	0.0248	0.0294**	−0.0096	0.0027	
Mountainous areas	−0.8361***	−0.7915***	0.1575	0.1475	−0.8807***	−0.8872***	
Plain areas	0.4399***	0.4327***	1.4616***	1.3992***	1.2169***	1.1729***	
_cons	0.4946	0.0149	−0.3081	−0.0586	−4.3450***	−4.3712	

Note: The missing dummy for region is hilly area. Significant level: *10%, **5%, ***1%. Number of cases=2777. Number of observations=11108. Log likelihood=−3538.5821/−3493.4122. Subscript (a) denotes model without household head characteristics; (b) denotes model with household head characteristics.

3.6 Conclusion

It is known that a switch from traditional biomass energy use to modern clean, safe, and efficient energy use could improve the local rural livelihoods by enhancing the access to high-quality energy and reducing the negative impacts of the traditional biomass energy on health, environment, and living standards. Hence, in this chapter, we examine the energy choice behaviors of the rural households in Sichuan Province. We found that the fuel switching in our study region occurs in the way that households will move up to the advanced energy (electricity) from the biomass energy with an improvement in their income levels or a decrease in electricity price. The common situation of multiple fuel use and the complicated process of fuel substitution indicate that the classic theory of the energy ladder is not appropriate to be applied to explain the household energy choice behaviors. Despite the use of other types of energy for cooking, merely 22.3% of households abandoned the use of traditional solid biomass energy. Meanwhile, the modern and clean fuels such as biogas, LPG, and natural gas are more popular in areas with higher income levels. This suggests that the decline in traditional biomass energy use is quite slow, and the energy transition is principally pulled by raising the household economic level.

Regarding the determinants of household energy choice behaviors, we categorized them into two groups, energy-specific and household-specific factors, and tested them respectively using revealed preference data, stated preference data, and joint RP-SP data. Our empirical work provides insights into these determinants based on observed energy adoption behaviors and hypothetical energy choices. It also has helped to explain why a large number of rural households in Sichuan Province are still unwilling to abandon the use of traditional solid biomass energy. The results indicate that the households prefer to use the types of energy with lower cost, higher safety, and lower indoor pollution. More importantly, this study also shows that the characteristics of the decision maker such as the age and educational level, the demographic structure of rural families such as the fractions of elderly people, children and adults, income level, arable land owned, number of people usually eating at home, distance to the nearest biomass collecting spot, and household locations are all crucial factors that affect the process of household energy transition.

Based on these findings, three suggestions are provided regarding the future policy design for further development of energy infrastructure construction and acceleration of energy transition in rural China. Firstly, the energy policies should focus more on the combined effect on both price and quality of modern fuels. Precisely, any sustainable energy transition policy should provide incentives that reduce energy price (or usage cost) while enhancing energy quality at the same time. For example, the most effective way for the promotion of biogas use is to advance modern technologies to shorten the time spent on operating and cleaning the digester and to improve the safety for the users. Secondly, the regional differences should be taken into account. Different regions have different situations. Therefore, the proposed energy policies have to be adapted to local conditions. In the mountainous areas where the production of biogas remains unsuitable, the policies should concentrate on how to provide electricity to the households with lower price and outage frequency, whereas in the plain and hilly areas, energy policy should provide a simultaneous promotion of biogas and electricity. Thirdly, indirect policy that enhances rural income and improves educational level should also be given more emphasis in policy design. On one hand, more skill-trainings related to the operation and maintenance of the biogas digesters should be provided. On the other hand, more opportunities for higher education and job outside the villages should be made for rural households.

Chapter 4 Evaluating the Impacts of Biomass Collection on Agricultural Production

4.1 Introduction

As a traditional agricultural country in the world, a large number of rural households are still living on agricultural production in China. Crops straw and firewood are the principal biomass resources generated in rural areas for energy use, as they are closely related to the main economic activity, namely agricultural production, which accounts for a significant portion of the productive activities across the whole country (Li et al., 2001). According to the statistics (MOA, 2010), the theoretical resources amount (TRA) of crop straw with 15% water content is 0.82 billion tons, while the available resources amount (ARA) of that is about 687 million tons, including 265 million tons of maize straws, 205 million tons of rice straw, and 150 million tons of wheat straw. Moreover, in total, 0.155 billion tons of woody biomass resources was derived from deforestation waste, and wood processing and firewood forests were used as feedstock for energy production by the end of 2010 (CRES, 2011). Despite this, due to lack of the access to modern conversion technologies such as biomass gasification, biomass briquetting, and co-combustion of coal and biomass, households usually collect crop straw after harvesting and pick up firewood on the way to or from the field, then directly burn them for domestic cooking or heating. Under this situation, biomass collection creates a burden on households that decide to use crops straw and firewood, which reduces the available labor inputs for agricultural production (Li et al., 2001; van der Kroon et al., 2013). Therefore, the issue of the impacts of biomass collection on agricultural production needs to be clarified and understood, including the relationship between these two productive activities and the effects of labor allocation.

Biomass collection, in our study regions, involves operations of gathering, packaging, and transporting biomass to a specific site (in most cases, households place the collected biomass near their houses) for temporary storage (Zafar, 2015). The amount of biomass resources that can be collected at a given time depends on a variety of influencing factors, such as geographical and seasonal variations in biomass, household labor allocation, local regulations on storage and transport, and so on (Sokhansanj et al., 2003). In the case of agricultural residues, large quantities of waste resulting from crops cultivation activities are promising objects for biomass collection activities. Furthermore, the access to biomass resources, the sequence of collection operations, and the efficiency of collection equipment are also important factors affecting biomass collection. Although the relationship between agricultural production and biomass collection has been investigated in many past studies (Li et al., 2001; Zhang et al., 2010; van der Kroon et al., 2013), the understanding and the empirical evidence of the impacts of biomass collection on agricultural production are still limited. Hence, the main purpose of this research is to examine how biomass collection affects agricultural production by modeling production with a focus on the allocation of labor inputs.

In accordance with the theoretical framework outlined in Section 2.3, the basic hypothesis of this research is that agricultural production competes with biomass collection for labor resources. Based on the economic theory of duality, we propose to test the hypothesis through investigating the product supply and input demand relationships. Here, we assume that households in our study region (in Sichuan Province) are clearly price taker and profit maximizing and competitive producers.

The structure of this chapter is organized as follows: previous works of literature are reviewed in Section 4.2. Section 4.3 presents a preliminary data analysis on the relationship between biomass collection and agricultural production. In Section 4.4, we describe the empirical strategies of this research, including model specification and estimation methods as well as the solutions to endogenous problems and imposition of restrictions. The estimation results of the models are given in Section 4.5, and our main findings are discussed in Section 4.6.

4.2 Literature Review

In recent years, works of literature directly concerning the impacts of biomass collection on agricultural production at household level are rare. Nonetheless, a large number of studies have focused on the effects of biofuel utilization on agricultural production (von Lampe, 2007; Baier et al., 2009; Kgathi and Mfundisi, 2009; Timilsina et al., 2010; Havey and Pilgrim, 2011; Ajanovic, 2011; Babcock, 2011; Zilberman et al., 2013; Alka et al., 2014). Most of them are works of qualitative analysis. The results of those studies demonstrate that the utilization of biofuel impacts agricultural production, both directly and indirectly. The direct influences come from the competition between energy crop cultivation and agricultural production for resources such as land and water (von Lampe, 2007; Baier et al., 2009), whereas the indirect effects are related to the mechanism of price transmission between biofuel prices and food prices (Havey and Pilgrim, 2011; Ajanovic, 2011; Zilberman et al., 2013).

With respect to the impact of biomass collection on agricultural production, many studies have empirically investigated the influence of fuelwood collection on household agricultural production (Kgathi, 1997; Heltberg et al., 2000; Fisher et al., 2005; Chen et al., 2006). As suggested by van Horen and Eberhard (1995), an increase in labor time spent on firewood collection may adversely influence the labor budget and, in turn, negatively affect agricultural production. That is to say, due to the limited time endowment, household members, especially women and children, who have to spend extensive amounts of time on firewood collection, are usually constrained from engaging in other income generating activities such as working off-farm and agricultural production (Li et al., 2011; van der Kroon et al., 2013). Additionally, from the perspective of environmental impacts, Mathye (2002) has pointed out that moderate amount of crop residues left in the field could not only prevent soil erosion, but also enrich the soil. Excessive crop straw collection will bring negative effects to agricultural production through soil depletion.

Despite the various efforts of previous works of research, the relationship between agricultural production and biomass collection and how biomass collection affects agricutultural production are issues that are still poorly understood. The empirical evidence from microeconomic studies is insufficient to clarify this issue. Therefore, this research tries to fill the gaps in existing literatures and focuses on household labor allocation decisions on different activities from the perspectives of household profit maximization problem.

4.3 Biomass Collection and Its Impacts on Agricultural Production in Study Region

4.3.1 Research Context in Sichuan Province

As a large traditional agricultural province in China, Sichuan has abundant biomass resources generated from agriculture and its related production activities. The main types of biomass collected by rural households for energy use are crop straw and firewood. Due to the different local geographic and weather conditions, households from different regions have different biomass resources and collecting habits. In mountainous areas, households often log in nearby forests in winter and transport the tree trunks to their home using tractors. After being cut down into small pieces, the firewood is piled up alongside the houses and finally used for cooking and heating during the subsequent year (See Figure 4.1(a)). As corn is the main kind of crop cultivated in mountainous areas, households living there also pick up its straws from their fields after harvesting. On the other side, in hilly and plain areas, crop straw plays a principal role in household biomass energy use. Particularly, the types of straw are determined by the types of crops planted in field. Therefore, both rice and wheat straw are adopted by households in plain areas for energy use, whereas rice and corn straw are collected in hilly areas. Additionally, for the households from hilly areas, bamboo trunks are usually used as firewood, while in plain areas, households prefer to collect fruit tree branches for cooking. In most cases, households employ tricycles to transport the collected biomass and let them lean against the wall of their houses (See Figure 4.1(b) and (c)).

Figure 4.1 Biomass collected in three different regions

(a) (b)

(c)

Note: a. firewood in mountainous areas; b. bamboo trunks and crops straw in hilly areas; c. fruit tree branches and crops straw in plain areas

4.3.2 Biomass Collection and Agricultural Production: Preliminary Data Analysis

In our study sample, 539 (96.94%) of the surveyed households participate in either agricultural production or biomass collection. The general information of household working activities is listed in Table 4.1. For the whole sample, with respect to household participation in different activities, 524 households engage in agricultural production, accounting for 94.24% of the total, while 409 households collect biomass, occupying 73.56% of all those surveyed. On average, the value of annual agricultural outputs is 17,208 CNY, whereas the amount of biomass collected by households is 3635 Kgsce per year. Moreover, the average time allocated to agricultural production is 717 hours per year, while the average time spent on collecting biomass is 236 hours annually. Regarding to agricultural production, households from Yibin (hilly areas) have the highest participation rate (96.23%), whereas households located in Deyang (plain areas) have the lowest one (91.89%). However, households from Aba (mountainous areas) spent the longest time (796 hours per year) on farm work, while those from Deyang spent the shortest time (613 hours per year). Accordingly, the total value of agricultural outputs in Aba is the largest (21678 CNY per year), whilst that in Deyang is the smallest (14329 CNY per year).

Turning to biomass collection, the participation rate differs among different areas. In Aba, the participation rate is the highest (92.42%), and it took the longest time (380 hours per year) for households to collect the largest amount of biomass (5354 kgsce per year). In contrast, Deyang has the lowest participation rate (45.95%), resulting in the fact that the time (74 hours per year) allocated to biomass collection is the shortest, and the amount of biomass collected by households is the smallest.

Table 4.1 *General information of household participation in agricultural production and biomass collection in study region*

	Aba (Mountainous areas)		Yibin (Hilly areas)		Deyang (Plain areas)		Total sample	
Household participation in two activities	Units	Share (%)	Units	Share (%)	Units	Share (%)	Units	Share (%)
Households participating in agricultural production	175	94.59	179	96.23	170	91.89	524	94.24
Households participating in biomass collection	171	92.42	153	82.26	85	45.95	409	73.56
Households participating in neither activities	3	1.62	2	1.08	12	6.49	17	3.06

	Aba (Mountainous areas)	Yibin (Hilly areas)	Deyang (Plain areas)	Total sample
Summary of household working activities (per household)	Mean value	Mean value	Mean value	Mean value
Value of agricultural outputs (CNY per year)	21678	15625	14329	17208
Amount of collected biomass (kgsce per year)	5354	4515	1031	3635
Hours worked on agricultural production (per year)	796	742	613	717
Hours worked on biomass collection (per year)	380	255	74	236

Source: Author's Own Household Survey (2013)

To be more specific, as shown in Table 4.2, of the 409 households that engaged in biomass collection, 313 households picked up crop straw in their fields after harvesting, while 347 households collected firewood. On average in each household, 1.46 members participated in collecting biomass accounting for about 39% of the total household size. Comparison of different regions reveals that the total number of households that collected biomass in Aba is the largest (171), whereas it is the smallest in Deyang (85). Aba has the largest number of households (165) collecting firewood, while Deyang has the smallest (70). Yibin has the largest number of households participating in biomass collection (150), while Aba has the smallest (78). In addition, households in Aba have the largest mean number of members (1.52) participating in biomass collection, while households from Deyang have the smallest (1.36). The proportion of members participating in biomass collection is the highest (44%) in households from Deyang, while it is the lowest (36%) among those living in Aba.

Table 4.2 General information of households collecting biomass

	Aba (Mountainous areas)	Deyang (Plain areas)	Yibin (Hilly areas)	Total Sample
Total number of households collecting biomass	171	85	153	409
crops straw	78	85	150	313
firewood	165	70	112	347
Average number of members participating in biomass collection	1.52	1.36	1.46	1.46
Proportion of members participating in biomass collection	0.36	0.44	0.39	0.39

Source: Author's Own Household Survey (2013)

Furthermore, the descriptive information of the 509 household members who are mainly responsible for biomass colletion is listed in Table 4.3. It can be seen that 63.85% of the household members are male and 40.86% of them are graduates of primary school. Adult members from the age group 45–54 represent the largest share (28.29%) of the total, while children (below 14 years old) occupy the smallest (0.2%). Considering the different regions, most of the household members in Aba are from age group 35–44, taking up 33.33% of the total. The majority of Deyang household members are elderly people (above 65 years old) accounting for 42.31% of the total, while the age group 55–64 takes the largest proportion (34.74%) among Yibin household members. Moreover, members with educational level of primary school occupy the largest share of 35.94%, 40.39%, and 45.54% in Aba, Deyang, and Yibin, respectively. And the male members take the dominating position in collecting biomass, accounting for 76.56%, 68.27%, and 50.23% of the total from the three regions. Finally, according to the statistics in Table 4.3, 89% of these households participated in agricultural production at the same time. Based on what has been described above, we can get to know that adult male members are the main labor force for biomass collection in our study regions. Importantly, household members who work on the farm also have to be responsible for biomass collection. That is to say, household labor allocation decision could impact the relationship between biomass collection and agricultural production.

Table 4.3 Descriptive information of household members who are mainly responsible for biomass collection

		Aba (Mountainous areas)		Deyang (Plain areas)		Yibin (Hilly areas)		Total Sample	
		Number	Share(%)	Number	Share(%)	Number	Share(%)	Number	Share(%)
Age	0–14	0	0	0	0	1	0.47	1	0.20
	15–24	5	2.60	0	0	0	0	5	0.98
	25–34	18	9.38	4	3.85	5	2.35	27	5.30
	35–44	64	33.33	5	4.81	30	14.08	99	19.45
	45–54	60	31.25	18	17.31	66	30.99	144	28.29
	55–64	32	16.67	33	31.73	74	34.74	139	27.31
	65+	13	6.77	44	42.31	37	17.37	94	18.47
Educational level	Illiterate or literate without formal schooling	34	17.71	19	18.27	39	18.31	92	18.07
	Literate below primary	29	15.10	5	4.81	26	12.21	60	11.79
	Primary school	69	35.94	42	40.39	97	45.54	208	40.86
	Middle school	49	25.52	33	31.73	39	18.31	121	23.77
	High school	10	5.21	5	4.81	11	5.16	26	5.11
	Diploma/certificate	0	0	0	0	1	0.47	1	0.20
	University/College graduate	1	0.52	0	0	0	0	1	0.20
Gender	Male	147	76.56	71	68.27	107	50.23	325	63.85
	Female	45	23.44	33	31.73	106	49.77	184	36.15
Working status	working on farm	166	86.46	93	89.42	194	91.08	453	89.00
Total		192		104		213		509	

Source: Author's Own Household Survey (2013)

As our interest lies in the relationship between biomass collection and agricultural production, we focus on the household labor allocation to these two intrahousehold production activities. In accordance with the basic hypothesis, households spent more time on biomass collection, implying that less time can be allocated to agricultural production, given a fixed amount of time for these two activities. Thus, the agricultural outputs decrease as a result of the decrease in labor inputs. In order to test this hypothesis, the relationship between agricultural production and biomass collection is illustrated in Figure 4.2 by firstly making a simple regression of the total value of agricultural outputs on the share of the time allocated to biomass collection. It can be seen from the line of fitted plots that, with an increase in the proportion of time spent on biomass collection, the total value of agricultural outputs decrease, accordingly. To some extent, this means that biomass collection has a negative effect on agricultural production.

Figure 4.2 The relationship between biomass collection and agricultural production

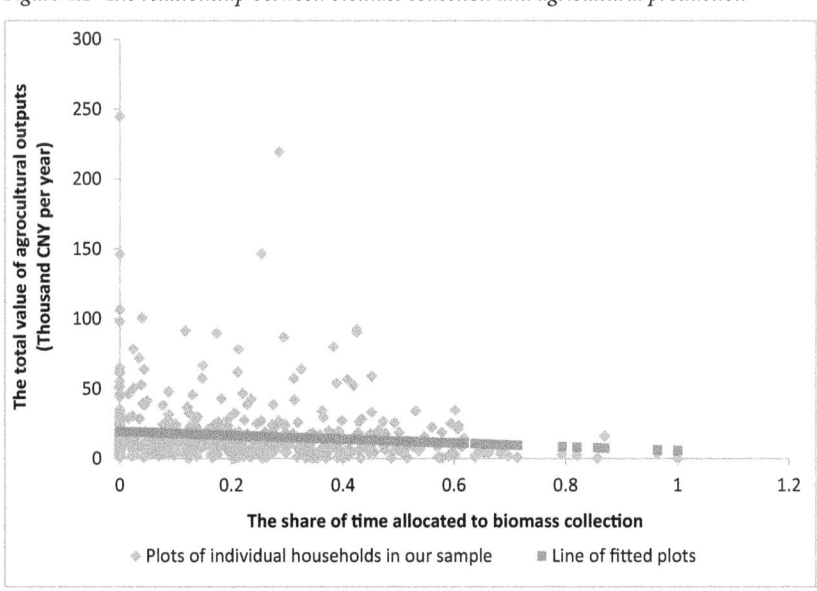

Source: Author's own field household survey

Once we have a preliminarily knowledge of the relationship between biomass collection and agricultural production, in the following sections, we propose to conduct an empirical analysis to further investigate the interlinkages between these two behaviors.

4.4 Empirical Analysis

In order to address our research question, a two-stage estimation strategy developed on the basis of previous literature (Henning and Henningsen, 2007a and 2007b; Tiberti and Tiberti, 2015) is adopted. We firstly estimate the shadow wage of household labor through modeling the production systems and then include the estimated shadow wage in a multi-output profit function to investigate the relationship between agricultural production and biomass collection.

4.4.1 Estimating Shadow Wage and Shadow Price

4.4.1.1 Econometric Specification

In the first step, the econometric specification of shadow wage estimation consists of household labor participation decision equations and a system of production functions.

- **Household Participation Decisions on Biomass Collection, Agricultural Production and Off-farm Work**

In order to investigate how household i makes decisions regarding participation in biomass collection, agricultural production, and off-farm work, we develop our econometric model on the basis of the reduced form of the labor allocation functions (2.18) in Section 2.3:

$$Y_{ni}^* = L_i^*(w_i, B_i, Z_i, T(a)_i, E_i) \; (n = a, b, o) \tag{4.1}$$

Where a denotes agricultural production, b indicates biomass collection, and o represents off-farm employment.

Then, we consider the first-order Taylor series expansion for labor allocation functions (4.1):

$$Y_{ni} = \alpha_{n0} + (\frac{\partial L_{ni}}{\partial w_i})w_i + \sum_{j=1}^{J}(\frac{\partial L_{ni}}{\partial X_{ij}})X_{ij} + \varepsilon_{ni} \tag{4.2}$$

Where ε_{ni} is the error term. $X = [X_1, \ldots X_j]$ represents explanatory variables other than market wage rate. Let us denote $\alpha_{n1} = \partial Y_{ni} / \partial w_i$ and $\alpha_{nj} = \partial Y_{ni} / \partial Y_{ij}$, and then we can create three estimable participation equations in the form as:

$$Y_{ni} = \alpha_{n0} + \alpha_{n1} w_i + \alpha_{nj} X_{ij} + \varepsilon_{ni} \quad (n = a, b, o) \tag{4.3}$$

It can be seen from (4.3) that the dependent variables are decisions on participation in one of the three activities ($Y_{ni} = 1$ if household participates in one activity, otherwise $Y_{ni} = 0$).

Among independent variables, w_i is the market wage rate. Theoretically, the functions in (4.1) reveal that market rate is one of the greatest determinants for household participation decision on off-farm work. When the market rate increases, households are more likely to participate in off-farm work. With respect for other explanatory variables in X, the other income E_i is measured by non-labor income. In our study, it mainly consists of subsidies provided from the government (such as subsidies for superior crop varieties, direct subsidies to grain cultivation, and subsidies for pig breeding, etc.), reimbursements from various insurances, remittances, and other returns from investment in the capital market. The household size and demographic characteristics (such as the fractions of children and elderly people) are used as a proxy for its time endowment $T(a)_i$. Households of larger size could have more labor resources, while the fractions of children and elderly people also reflect the amount of available labor resources that can be provided by the household. Normally, households with larger fractions of children and elderly people are less likely to allocate labor to off-farm work. For the other inputs in agricultural production (B_i), we firstly choose the areas of arable land owned by the household, as it can significantly influence the household decision regarding participation in agricultural production. We do not use the total value of intermediate inputs as an explanatory variable in our econometric model, because household allocation decision can affect intermediate inputs using activities, especially the use of fertilizers and pesticides, according to the findings of many previous works of research (Lamb, 2001; Mathenge and Tschirley, 2007). Instead, we use the weighted price of fertilizers and pesticides (calculated using formula (4.11)) as a proxy for the amount of intermediate inputs, since the households are more likely to purchase cheaper fertilizers and pesticides. Meanwhile, considering biomass collection, the distance from house to the nearest biomass collecting spot is selected to represent accessibility and availability. It is also supposed to negatively influence household labor allocation to biomass collection. Particularly, for the omission of this variable caused by non-participation in biomass collection, we assume that these households face the average distance and substitute regional sample mean for the missing data.

We also include household head characteristics such as age, gender, and educational level in our model, because these characteristics affect the quality of household labor, and then influence the marginal products of land and other intermediate inputs, which in turn brings effects to household participation decisions on different activities. Finally, the household location dummies are added into the regression to capture the effects of regions on household labor participation decisions.

In order to estimate the participation equations, we employ a multivariate probit model and then use the method of simulated maximum likelihood (SML) to obtain the estimate results of the model.

- **Household Production System**

As it has been discussed in the theoretical section, the shadow wage determines the household decision on labor allocation, which in turns affects household behaviors of agricultural production and biomass collection. Under this circumstance, we firstly estimate the shadow wage of the household's labor using production function. According to the multioutput production function obtained using the implicit function theorem in (2.9) and considering the easiness of estimation and interpretation, the simultaneous agriculture-energy production relationship for household i can be represented by a system of two equations derived from the Cobb-Douglas transformation function as follows (Just et al., 1983; Debertin, 2012):

$$\ln TOA_i = \sigma_0 + \sigma_1 \ln TOB_i + \sigma_2 \ln L_{ai} + \sum \sigma_m \ln B_{mi} + \sum \sigma_k d_{ki} + v_i \qquad (4.4)$$

$$\ln TOB_i = \lambda_0 + \lambda_1 \ln TOA_i + \lambda_2 \ln L_{bi} + \sum \lambda_j d_{ji} + \mu_i \qquad (4.5)$$

Where μ_i and v_i are the error term.

In this model, the agricultural output (variable name *TOA*, measured by the total value of agricultural products, i.e. the quantities of crops produced by household i multiply the prices of these crops) is modeled as a function of the quantity of biomass (denoted as *TOB*, measured by the total amount of the biomass collected by household i, since the biomass or biomass energy is nontradable. In order to unify the units of firewood and crop straw to standard coal equivalent (kgsce), we divide the quantities of them by their conversion coefficients[12] respectively). The quantity of labor input (L_a, the total hours worked on farm), a vector of other inputs (B_m, including the areas of arable land *AL*, and the total value of intermediate inputs *TCI*. Due to the unavailability of the data, we use the total cost of fertilizer, pesticides and plastic films instead), and other variables (d_k), such as household local dummies (r_1 and r_2) that can also influence households' agricultural production. In contrast, the amount of collected biomass is hypothesized as a function of the total value of agricultural outputs, the labor input (L_b, the time spent on biomass collection), and other influencing

---

12 The data of conversion coefficients for all types of energy are collected from China Energy Statistic Yearbook (2009).

factors d_j (including the distance to biomass collecting spots DB and household location dummies).

Once the system of equations (4.4)–(4.5) has been estimated, the shadow wage of household labor can be calculated using the following formula:

$$w_i^* = MPL_{ai} = \frac{\hat{\sigma}_2 \hat{TOA}_i}{(1-\hat{\sigma}_1\hat{\lambda}_1)L_{ai}} \tag{4.6}$$

Where $\hat{TOA}_i$ is the predicted value of agricultural output and $\hat{\sigma}_1, \hat{\sigma}_2$, and $\hat{\lambda}_1$ are the estimated coefficients associated with outputs and labor allocated in agricultural production, respectively.

As the market for biomass energy is almost absent in our study region, the prices of biomass energy (crop straw and firewood) cannot be directly observed. Therefore, according to the equilibrium condition $MPL_{ai} = MPL_{bi} = w_i^*$, we use the shadow wage derived from (4.3) to calculate the shadow prices of crop straw and firewood as:

$$Shadow\ price = \frac{w_i^*(CNY\ per\ hour) \times Collecting\ time\ (Hours)}{Total\ amount\ of\ collected\ biomass\ (kg)} \tag{4.7}$$

The calculated shadow wage rates and shadow prices will be included in estimating the profit function in the next step (See Section 4.4.2 for details).

The Ordinary Least Squares (OLS) estimates of the production system may be biased for three main reasons. Firstly, unobserved information such as the ability and management level of the household reflected in the error terms are likely to be correlated with the endogenous variables, particularly the variable inputs (labor and intermediate inputs) in both of the equations, which may lead to omitted variable bias. Secondly, the two output variables are jointly determined. Thus, the single-equation estimation may suffer from simultaneity bias, due to the correlation between the disturbance of each equation and the output variables. Moreover, since the output variables are also the dependent variables of the equations in the system, the error terms among the equations are also expected to be correlated (Greene, 2012), which may cause the problem of inefficient estimation. Thirdly, the observed data can only reflect the situation of the households who decide to participate in corresponding production activities. Under this circumstance, the conditional means of error terms over the non-zero output population are not equal to zero, implying that the potential sample selection bias should be corrected in our model estimation.

The first problem could be mitigated by including observable household characteristics such as age, gender, and educational level as proxies for management ability for both of the production activities. Finally, all variables used to estimate the simultaneous production system are listed in Table 4.2.

The second problem is solved by using estimation methods for simultaneous equations. In this research, IT3SLS (iterative three-stage least squares) is applied to estimate the system of production functions. The IT3SLS method combines the procedure of the 2SLS (two-stage least squares) and SURE (Seemingly unrelated estimation) and produces the system estimates from a three-stage process (Zellner and Theil, 1962). In the first and second stages, an instrumental-variables approach is adopted to develop instrumented values for all endogenous variables (the output variables in the system) and to obtain a consistent estimate for the covariance matrix of the equation disturbances. All other exogenous variables in the system are used as instruments. In the third stage, generalized least squares (GLS) estimation is performed using the covariance matrix estimated before and with the instrumented values in place of the right-hand-side endogenous variables (Greene, 2012). AThen, this process iterates over the estimated disturbance covariance matrix and parameter estimates until the parameter estimates converge.

The third problem can be solved by the standard two-stage Heckman (1979) sample selection model. As shown in Table 4.1, in our sample, the proportion of non-zero observations in biomass collection is only about 73.56%. Therefore, using only the data of the households that decide to participate in biomass collection may cause sample selection bias. To deal with this problem, in the first stage, the results of the multivariate probit regression model, which is estimated to determine the probabilities that a given household will participate in agricultural production and biomass collection, are used to calculate the Inverse Mills Ratio (IMR) for each household. In the second stage, parameter estimates of the production system are obtained by augmenting the regression with the IMR using 3SLS (Heckman, 1979). Based on the equations in (4.1) and (4.3) presented in the last step, the IMRs for household i who chooses to participate in either activity can be computed as follows:

$$IMR = \phi(w_i, X_{i,j}, r1, r2) \big/ \Phi(w_i, X_{i,j}, r1, r2) \tag{4.8}$$

Additionally, to deal with the zero-value variables that have undefined logarithm, we modify them by replacing them with a "sufficiently small" value (MaCurdy and Pencavel, 1986; Jacoby, 1992; Soloaga, 1999).

4.4.1.2 Data Description

The variables used in estimating the production systems are listed in Table 4.4. These variables can be categorized into household head characteristics, household demographic characteristics, household productive characteristics, and other external factors.

In terms of household head characteristics, the average age of the heads of the surveryed households is 51.74 years, and their average schooling year is 6.42 years. The share of the male household heads in the total sample is about 0.93. For households from different regions, these characteristics are different. In Aba, household heads have the lowest average age (47.91) and the lowest average educational level (5.92). The share of the male household heads there is the smallest (89%). In contrast, Deyang has the highest mean values of the age (54.03) and the educational year (7.02) of the household heads, while Yibin has the largest share of male household heads (97%).

For household demographic characteristics, the average family size of the households in our sample is 4.12. The mean values of the fractions of children and elderly people in household members are 0.11 and 0.12, respectively. Households from Aba have the largest size (4.39) and the smallest fraction of elderly people (0.06), whereas households located in Deyang have the smallest size (3.62), the smallest fraction of children (0.06), and the largest fraction of elderly people (0.19). Moreover, the fraction of children (0.14) is the largest in Yibin.

Table 4.4 Description of household characteristics and variables used in model estimation[13]

Variable name	Aba		Yibin		Deyang		Total sample	
	Mean	Std. Dev.	Mean	Std. Dev.	Mean	Std. Dev.	Mean	Std. Dev.
Age of household head (years)	47.91	10.55	53.27	10.76	54.03	12.31	51.74	11.54
Age squared of household head	2406.37	1053.17	2953.19	1203.16	3070.34	1365.23	2810.22	1245.77

---

13 The other four variables (value of agricultural outputs, amount of collected biomass, hours worked on farm, and hours worked on biomass collection) are described in Table 4.1.

Variable name	Aba		Yibin		Deyang		Total sample	
	Mean	Std. Dev.	Mean	Std. Dev.	Mean	Std. Dev.	Mean	Std. Dev.
Gender of household head (share of male)	0.89	0.32	0.97	0.16	0.92	0.27	0.93	0.26
Educational level of household head (years)	5.92	3.56	6.33	3.55	7.02	3.25	6.42	3.48
Family size	4.39	1.29	4.33	1.52	3.62	1.13	4.12	1.37
Fraction of children (≤ 14)	0.13	0.18	0.14	0.16	0.06	0.10	0.11	0.16
Fraction of elderly people (≥ 65)	0.06	0.13	0.11	0.21	0.19	0.31	0.12	0.23
Arable land areas (Mu)	4.52	4.21	4.62	4.07	2.99	2.21	4.05	3.69
Weighted price of fertilizers and pesticides (CNY per kg)	8.85	9.43	3.75	3.02	6.23	4.18	6.27	6.53
Market wage rate (CNY per hour)	7.38	4.32	8.08	4.60	9.39	5.96	8.29	5.07
Non-labor income (CNY per year)	3375.96	4254.01	2275.62	4519.24	2719.23	4626.81	2789.01	4484.31
Distance to biomass collecting spot (km)	5.16	6.64	0.79	0.68	0.49	0.57	2.14	4.41

Source: Author's own household survey

Considering household productive characteristics, on average, the area of arable land possessed by the surveyed households is 4.05 Mu. Households in Aba own the largest mean areas of arable land, while households from Deyang have the smallest. The weighted price of fertilizers and pesticides is the highest (8.85 CNY per kg) in Aba, while being the lowest (6.23 CNY per kg) in Yibin. The mean value of this price index in our sample is 6.27 CNY per kg.

Among other external characteristics, the average market wage rate for the sampled households is 8.29 CNY per hour, and the average non-labor income level is 2789.01 CNY per year. More specifically, the average market wage rate is the highest (9.39 CNY per hour) in Deyang and the lowest (7.38 CNY per hour) in Aba. Households in Aba have the highest non-labor income level at 3375.96 CNY per year, whilst households in Yibin have the lowest non-labor income level of 2275.62CNY per year. Additionally, in our sample, the biomass collecting spot is on average 2.14 km away from the house. Due to the local conditions, Aba has the longest distance (5.16 km), whereas Deyang has the shortest (0.49 km).

4.4.1.3 Model Estimation Results

Table 4.5 lists the estimation results of the multivariate probit model that explains how households make decisions regarding participation in different activities. The estimates of ρs (Rho, correlation between the errors) that maximizes the multivariate probit likelihood function are 0.1095, −0.0114, and 0.0291, respectively. Specifically, the correlation coefficient between agricultural production and biomass collection (Rho (b, a)) is positive and significantly greater than zero at the level of 5%. This indicates that the random disturbances in participation equations of agricultural production and biomass collection are affected in the same direction by random shocks. In other words, household participation decisions on these two activities are not statistically independent. The correlation coefficients between off-farm work and the two intrahousehold production activities are insignificant, implying that the household participation decision regarding off-farm work does not statistically depend on participation decisions about intrahousehold production activities. This also means that we can separately analyze the relationship between biomass collection and agricultural production without considering off-farm labor allocation (See 4.4.2 for specification of multioutput SNQ profit function).

The significant log pseudo likelihood statistic suggests that the independent variables taken together influence household participation decisions. According to the estimated parameters, households that have a higher wage rate are more likely to participate in off-farm work and less likely to collect biomass. These results are in line with our expectation.

Table 4.5 Multivariate Probit Estimates of household participation functions of agricultural production, biomass collection and off-farm work

Variable	Agricultural production (D_{ai})		Biomass collection (D_{bi})		Off-farm work (D_{oi})	
	Coefficient	Std.Dev.	Coefficient	Std.Dev.	Coefficient	Std.Dev.
Age of household head	−0.1453**	0.0723	0.0531	0.0485	−0.0248	0.0441
Age squared of household head	0.0014**	0.0007	−0.0004	0.0005	0.0002	0.0004
Gender of household head	−0.0938	0.3472	−0.4138	0.2773	0.2412	0.2724
Educational level of household head	−0.0072	0.0303	−0.0458**	0.0219	−0.0041	0.0275
Family size	0.0799	0.0719	0.0643	0.0598	0.4365***	0.0747
Fraction of children (≤14)	−0.6252	0.6014	0.3226	0.5366	−1.1832**	0.5279
Fraction of elderly people (≥65)	0.1732	0.4400	0.6679	0.4086	−1.1086***	0.3436
Arable land areas	0.0789	0.0491	0.0067	0.0242	−0.0323*	0.0184
Market wage rate (log)	−0.1129	0.1419	−0.2182**	0.1066	0.1434*	0.0742
Non-labor income (log)	−0.2085**	0.0807	0.0947	0.0581	−0.1292*	0.0665
Price index of fertilizers and pesticides (log)	−0.1144	0.0920	0.1431	0.1039	−0.2165*	0.1105
Distance to biomass collecting spot	0.0547	0.0569	0.6374***	0.1179	0.0298	0.0354
Mountainous areas	−0.0859	0.3128	−0.1299	0.2436	−0.2744	0.2453
Plain areas	−0.1572	0.2438	−1.0145***	0.1755	−0.0638	0.2047
_cons	6.8066***	2.1738	−1.3185	1.4729	1.4401	1.3916
Log pseudolikelihood	−496.59624					
Rho (b,a)	0.1905**	0.0885				
Rho (o,a)	−0.0114	0.1225				

Variable	Agricultural production (D_{ai})		Biomass collection (D_{bi})		Off-farm work (D_{oi})	
	Coefficient	Std.Dev.	Coefficient	Std.Dev.	Coefficient	Std.Dev.
Rho (o,b)	0.0291	0.0951				
Wald chi2 (42)	228.00***					
No. of Obs	556					

Note: The significance levels are *10%, **5%, and ***1%. The missing location dummy is hilly area

As expected, non-labor income has a significant impact on household participation decision regarding agricultural production and off-farm work. Households with higher no-labor income level are less likely to allocate time to on-farm work and off-farm work. The price index of intermediate inputs has a significant negative impact on off-farm work. That means raising the intermediate input price reduces household likelihood to participate in off-farm employment. The areas of arable land owned by households can also significantly influence household participation in off-farm work. Houesholds possessing more arable land are less likely to find jobs outside their farms.

In terms of the demographic characteristics, households with larger size and smaller fractions of children and elderly people are more likely to work off-farm. With respect to the household head characteristics, we find that households with older heads are less likely to work on farm, whereas households with higher educational level are less likely to collect biomass. In addition, household location plays a vital role in determining household participation in biomass collection. Households located in plain areas are less likely to participate in biomass collection than those located in hilly areas.

Table 4.6 presents the iterative three-stage least squared (IT3SLS) estimates of the production system. The R^2 for the two equations are 0.3677 and 0.2835. The inverse Mills ratio *IMR* is insignificant in both equations, indicating that sample selection bias would not happen if the system of production functions was estimated without taking household participation decisions on biomass collection into consideration.

Table 4.6 Estimation results of the system of production functions using IT3SLS

Variable	Agricultural production		Biomass collection	
	Coefficient	Std. Dev.	Coefficient	Std. Dev.
Total value of agricultural outputs (log)			0.0524	0.0803
Amount of collected biomass (log)	−0.0597	0.1232		
Hours worked on farm (log)	0.6241***	0.0935		
Hours worked on biomass collection (log)			0.3634***	0.0464
Total value of intermediate inputs (log)	0.0171	0.0146		
Areable land areas (log)	0.2763***	0.0810		
Age of household head	0.0360	0.0288	0.0116	0.0303
Age squared of household head	−0.0004	0.0003	−0.0002	0.0003
Gender of household head	0.0790	0.1771	0.1280	0.1803
Educational level of household head	0.0488***	0.0150	0.0182	0.0154
Distance to biomass collecting spots			0.0083	0.0123
Mountainous areas	0.3989***	0.1203	−0.1524	0.1381
Plain areas	0.2810	0.1791	−0.4265**	0.1806
IMR	0.3777	0.2778	−0.4154	0.2697
_cons	3.5683***	1.4237	5.5164***	0.9951
R^2	0.3677		0.2835	
No. of Obs.	394		394	
Endogenous variables[a]	ln_TOA, ln_TOB			

Note: The significance levels are *10%, **5%, and ***1%. The missing dummy for regions is hilly area. a. All other variables in this system are treated as exogenous to the system and uncorrelated with the disturbances. The exogenous variables are taken as instruments for the endogenous variables.

With respect to the parameters of the production system, most of them have the expected signs. For the agricultural production, the inputs of labor and arable land have significantly positive impacts on the outputs. The educational level of the household head has a significant effect on farm production, supporting the widely accepted role of human capital in improving agricultural production (Henning and Henningsen, 2007; Tiberti and Tiberti, 2015). In addition, households located in mountainous areas produce more agricultural products than those from hilly areas. On the other hand, in biomass collection function, the labor input also has a significant and positive influence. The estimated parameters also indicate that households who are from plain areas collect less biomass than those from hilly areas. In addition, the coefficients of the output variables on the right-hand side (RHS) of the two equations to some extent imply the relationship between agricultural production and biomass collection. Given the fixed labor inputs, spending more time on biomass collection decreases the outputs of agricultural production. Conversely raising the yields of agricultural production could also increase the collection amount of biomass. This could be possibly attributed to the fact that biomass collection occupies labor resources for agricultural production. Nonetheless, agricultural production provides biomass resources. Furthermore, due to the fact described in Section 2.3 that in Sichuan Province, household members usually collect biomass on the way to and from the fields, participating in agricultural production may increase the opportunity to pick up biomass.

After getting the parameter estimation results of the production system, the shadow wage of household labor and the shadow prices of the biomass energy are calculated using (4.3) and (4.4) (See Table 4.5 for the results of the calculation).

4.4.2 Estimating Profit Function

4.4.2.1 Econometric Specification

In order to further investigate the impacts of biomass collection on agricultural production, we estimate a multi-output profit function to obtain the full coefficients of the profit function as well as the price elasticities with respect to all outputs and inputs in the second step. Regarding the specification of the profit function, a number of plausible functional forms have been discussed in previous works of literature. They include the translog (TL), generalized Leontief (GL), normalized quadratic (NQ), symmetric normalized quadratic (SNQ), and many other forms (Christensen et al., 1973; Lau, 1972; 1978; Diewert and Wales, 1987; 1988; 1992; Diewert and Ostensoe, 1988; Kohli, 1993; Villezca-Becerra and Schumway, 1992). As described in Kohli (1993), the SNQ profit function treats all outputs and inputs symmetrically (NQ profit function can be considered as a special case of SNQ profit function). It is necessarily linearly homogeneous in prices and quantities, and as a fully flexible functional form, it can easily assume monotonicity and convexity properties (Kohli, 1993). Therefore, we adopt a symmetric normalized quadratic (SNQ) profit function defined as follows (Diewert and Wales, 1987, 1992; Henningsen, 2015):

$$\pi(p,z) = \sum_{i=1}^{n}\alpha_i p_i + \frac{1}{2}W^{-1}\sum_{i=1}^{n}\sum_{j=1}^{n}\beta_{ij}p_i p_j + \sum_{i=1}^{n}\sum_{j=1}^{m}\delta_{ij}p_i z_j + \frac{1}{2}\sum_{i=1}^{n}\sum_{j=1}^{m}\sum_{k=1}^{m}\gamma_{ijk}p_i z_j z_k \quad (4.9)$$

With π =profit, p_i =netput prices, z_i =quantities of non-allocable quasi-fixed inputs, $W = \sum_{i=1}^{n} \theta_i p_i$ =price index for normalization, θ_i =weights of prices for normalization, and α_i, β_{ij}, δ_{ij} and γ_{ijk} =coefficients to be estimated.

Given the above specification, the estimation equations (output supply and input demand equations) used to analyze household production decisions are obtained by the first derivation of the profit function using Hotelling's Lemma ($q_i = \partial \pi / \partial p_i$):

$$x_i = \alpha_i + W^{-1} \sum_{j=1}^{n} \beta_{ij} p_j - \frac{1}{2} \theta_i W^{-2} \sum_{j=1}^{n} \sum_{k=1}^{n} \beta_{jk} p_j p_k + \sum_{j=1}^{m} \delta_{ij} z_j + \frac{1}{2} \sum_{j=1}^{m} \sum_{k=1}^{m} \gamma_{ijk} z_j z_k \quad (4.10)$$

Where x_i =netput quantities.

In this research, we have four groups of netputs ($n = 4$), i.e. agricultural output (a), biomass output (b), labor input (l), and intermediate inputs (such as fertilizers and pesticides)[14] (o). Arable land is specified as the only quasi-fixed input. The price data were obtained from our field survey. Due to the fact that each group has many different individual output and input categories and the variations in the price of the same commodities are quite small, therefore, with the exception of labor input, within the other three groups, it is necessary to aggregate the price and quantity data of different individual outputs and inputs. In this study, we set up a household-specific price index by calculating the sum of weighted prices of each category using output value structure in each group. The price of each group of netput can be defined as follows (Lewbel, 1989):

$$p_i = \sum_{i=1}^{n} p_n s_n \quad (n = 1, 2 \cdots N) \quad (4.11)$$

Where s_n is the share of the value of netput n in netput group i and p_n is the producer price of netput n (as we do not have the price data of intermediate inputs for each household, we use the sum of weighted prices of fertilizer, pesticide, and plastic films which was calculated using the average price and consumption structure data of the sampled households instead). For the households that do not participate in either of the two productive activities, the corresponding production data are missing. We keep their output quantities zero and assume that these households face the average prices and replace the missing data with a sample mean. In particular, for agricultural output (x_{ai}) and intermediate inputs

---

14 In total, we set up a system of four equations (two outputs equations and two inputs equations).

(x_{oi}), the aggregated quantities are computed through dividing their total value by their weighted prices.

Moreover, we employ the following formula outlined by Diewert and Wales (1992) to calculate the weights θ_i:

$$\theta_i = \frac{|\bar{x}_i| p_i}{\sum_{i=1}^{n} |\bar{x}_i| p_i} \tag{4.12}$$

Once the SNQ profit function has been estimated, we define the price elasticity as:

$$E_{ij} = \frac{\frac{\partial q_i}{q_i}}{\frac{\partial p_j}{p_j}} = \frac{\partial q_i}{\partial p_j} \cdot \frac{p_j}{q_i} \tag{4.13}$$

Then, we will use these price elasticities to further analyze the relationship between agriculture production and biomass collection.

According to Microeconomic theory, we must consider the conditions imposed on our SNQ profit function before estimating it. Homogeneity in netput prices is imposed by the functional form, and symmetry requires $\beta_{ij} = \beta_{ji}$, $\forall i, j = 1, \cdots, n$ (Henning and Henningsen, 2007). In addition, in order to be consistent with the solutions to the profit maximization problem, the profit function has to be convex in netput prices (Varian, 1978). This implies that the Hessian matrix of the profit function must be positive semidefinite (Arnade and Kelch, 2007). Therefore, we applied the three-stage procedure proposed by Koebel et al. (2000; 2003) to impose convexity on the SNQ profit function. Firstly, we calculate the Hessian matrix after estimating the unrestricted netput equations in (4.6). Then we minimize the weighted difference between the unrestricted Hessian matrix and a Hessian matrix that is restricted as a positive semidefinite by the Cholesky factorization. In the last stage, we estimated the restricted coefficients by adopting an asymptotic least squared (ALS) framework (Gourieroux et al., 1985; Kodde et al., 1990; Henning and Henningsen, 2007).

As the shadow wage of household labor and the shadow price of biomass energy are unobservable and endogenously determined in the production system, an estimating process of instrumental variable regression should be employed in the estimation of the netput equations. Here, we choose the average age and education years of household working members as instrumental variables for the shadow wage (p_l^*) and shadow price (p_b^*). These variables can be assumed to be exogenous in our model, and the characteristics of the working members may affect the quality of household labor, despite the fact that the instruments are assumed to be less correlated with household labor allocation decision, and therefore are appropriate instrumental variables. In order to estimate our SNQ profit function, we firstly regress the shadow wage (p_l^*) and shadow price (p_b^*) on the instrumental variables and all the other exogenous variables. Then the predicted value of these two endogenous variables will be used as augmented variables in the constrained IT3SLS at the second step.

Using the iterative three-stage least square (IT3SLS) estimation method, we jointly estimate the SNQ profit function and the four netput equations with the data collected from Sichuan Province. As described already, restrictions are imposed on the system to ensure profit maximization. The estimations and calculations for the SNQ profit function are carried out by the statistical software "R" with the add-on package "micEconSNQP."

4.4.2.2 Data Description

The data used to estimate the SNQ profit function are presented in Table 4.7. On average, the total amount of agricultural outputs is 55562.34 kg per year, while the total amout of biomass collected by the sampled households is 3634.933 kgsce per year. The total time annually allocated to both agricultural production and biomass collection is 953.227 hours, whereas the quantity of intermediate inputs used per year is about 850.711 kg.

Table 4.7 Data and variables used in estimating SNQ profit function

	Description	Mean	Std. Dev.
x_a	Quantity index of agricultural outputs (kg)	55562.34	187089.9
x_b	Total amounts of collected biomass energy (kgsce)	3634.933	4464.771
x_l	Total hours allocated to two activities (Hours)	−953.227	655.108
x_o	Quantity index of Intermediate inputs (kg)	−850.711	886.112
p_a	Price index of agricultural products (CNY/kg)	1.269	1.926
p_b^*	Shadow price index of biomass (CNY/kgsce)	0.810	1.225
p_l^*	Shadow wage rate of household labor (CNY/hour)	10.749	4.933
p_o	Price index of intermediate inputs (CNY/unit)	7.337	22.786
z_{AL}	Cultivated arable land areas (Mu)	4.045	3.686
No. of Obs.			556

Source: Author's own field survey

With regard to the price indicies of netputs, the shadow wage rate of household labor is on average 10.749 CNY per hour, while the average shadow price index of biomass (crops straw and firewood) is 0.810 CNY per kgsce. The mean value of the price index of agricultural products in our sample is approximately 1.269 CNY per kg, whereas that of the price index of intermediate inputs is about 7.337 CNY per unit.

4.4.2.3 Model Estimation Results

Table A.1 reports the estimates of the SNQ production function with restriction of curvature. The hausman test statistic indicates that the endogeneity problem caused by including shadow wage in our model and our instrumental variables are not weak. Given our estimation results, we calculate the price elasticities of outputs and inputs according to (4.13) using sample means (See Table A.1 in the Appendix).

Table 4.8 Estimated price elasticities of outputs and inputs

	P_a	P_b^*	P_l^*	P_o
x_a	0.0420	−0.0219	−0.0067	−0.0133
x_b	−0.5247	0.2798	0.1212	0.1237
x_l	0.0463	−0.0348	−0.0755	0.0640
x_o	0.1504	−0.0584	0.1051	−0.1971

Note: The elasticities are calculated using R package micEconSNQP. The superscript u refers to the estimated coefficients of unrestricted profit function, whereas r is those of restricted estimation. T-Stat refers to the estimate parameter to the left. Subscript a represents agricultural outputs, b denotes amount of collected biomass, l is labor inputs, and o refers to intermediate inputs.

It can be seen from Table 4.8 that all outputs and inputs are inelastic. The own-price elasticities of outputs indicate that if the weighted average price of agricultural products increases 1%, the agricultural outputs will rise by about 0.04%, whereas a 1% increase in the shadow price of biomass energy will increase the outputs of biomass collection by about 0.28%. Meanwhile, the own-price elasticities of inputs also suggest that a 1% increase in the shadow wage of household labor will decrease labor input for the productive activities by 0.08%, whereas a 1% increase in the weighted average price of intermediate inputs will reduce household demand for them by 0.20%. Considering the cross-price elasticities, the supply (agricultural products and biomass energy) cross-price elasticities are negative, revealing a competitive relationship between these two activities. In other words, an increase in the price of either of the outputs leads more labor inputs to be invested in producing it, thus reducing the production of the other. This is also in line with the findings of our theoretical analysis in Section 2.3. Additionally, the cross-price elasticities for the inputs (labor and intermediate inputs) are positive, reflecting that the intermediate inputs such as fertilizers and pesticides are substitutes to labor-capital in our study region. This is to say, holding other variables constant, if the price of intermediate inputs increases, households will use less of them and simultaneously allocate more labor to production activities in order to keep the same quantities of outputs and vice versa. Moreover, if we compare the cross-price elasticities of intermediate inputs and labor (i.e. $\left|E_{x_o p_l^*}\right| > \left|E_{x_l p_o}\right|$), the labor-intensive feature of the production system in rural Sichuan Province is then confirmed. However, if we compare the own-price elasticities of the outputs with their cross-price elasticities, respectively (i.e. $\left|E_{x_a p_a}\right| > \left|E_{x_a p_b^*}\right|$ and $\left|E_{x_b p_a}\right| > \left|E_{x_b p_b^*}\right|$), it demonstrates that both agricultural production and biomass collection are more likely to be driven by the market of agricultural products than the demand of biomass energy. Particularly, for agricultural production, the negative signs of the cross-price elasticities of outputs with respect to inputs are consistent with economic theory. In contrast, although fertilizers and pesticides are not directly invested in biomass collection, the positive signs of the cross-price elasticities to inputs imply that biomass collection is perhaps influenced by consumption decisions. When the price of other inputs increases, households have to spend more on purchasing them and cut down their expenditures on commercial energy under a given budget constraint. As the consequence, they collect more biomass for energy use to compensate for the consumption of commercial energy. On the other side, if the shadow wage increases, households will work on domestic production activities for a longer duration instead of working off-farm, resulting in a decrease in their disposable incomes. Therefore, they have to use biomass as fuels to reduce the expense on commercial energy.

4.5 Conclusion

In this chapter, we analyze the impacts of biomass collection on agricultural production in our study region. This can be relevant in a context such as Sichuan and many other provinces in China, where the government has recognized the importance of biomass collection to environmental protection by reducing the pollution caused by directly burning agricultural wastes and firewood in open air.

Firstly, in terms of household participation in different working activities, the results of our study show that household participation decisions on agricultural production and biomass collection are not statistically independent. The educational level of the household head, market wage rate, non-labor income level, and household location are important factors in determining household participation in these two activities. Households that have well-educated heads and higher market wage rate are less likely to engage in both of these two productive activities. Households located in plain areas are less likely to work on farm as well as to collect biomass. In addition, households whose houses are farther from the biomass collecting spots are more likely to participate in both of these activities. In particular, non-labor income level is a key factor that can significantly influence household participation decisions on agricultural production and biomass collection in opposed directions. An increase in household non-labor income decreases the likelihood to work on farm, while increasing the household participation probability of biomass collection.

The results of our analysis also show that the supply cross-price elasticities of agricultural products and biomass energy are −0.02 and −0.52, respectively, indicating that biomass collection could bring negative impact to agricultural production due to the competition between these two activities for the limited labor resources. Whereas agricultural production provides residues as feedstock for biomass energy utilization, thus it could positively affect biomass collection.

Considering the changes in prices (shadow prices) of the outputs and inputs of both agricultural production and biomass collection, we found that the relationship between agricultural production and biomass collection is competitive. Biomass collection is likely to be driven by the markets of agricultural products and intermediate inputs. Increasing the price of agricultural products leads to more agricultural production and less biomass collection at the same time. Moreover, we also found that biomass collection is influenced by household consumption decisions. If the price of intermediate inputs increases, households will spend more on purchasing them and cut down their expenditures on commercial energy under a given budget constraint. Therefore, they have to collect more biomass for energy. On the other side, if the shadow wage increases, households will work on domestic production activities for a longer duration instead of working off-farm, resulting in a decrease in their disposable incomes. In this case, they also have to collect more biomass as fuels in order to reduce the expense on commercial energy. Based on the above research results, we suggest that measures aiming at promoting agricultural production, such as increasing the price of agricultural products while decreasing price of intermediate inputs, should be taken to reduce biomass collection. In addition, education in rural areas should also be developed to increase the probability of household members working off-farm.

Chapter 5 Impacts of the Changes in Exogenous Markets on Household Biomass Energy Use

5.1 Introduction

In the real world, an agricultural household usually plays a double role of producer and consumer in domestic biomass energy utilization. As a consumer, in order to meet its energy demand, the household must not only choose which types of energy it will use, but also decide the consumption amount of each type of energy on the basis of its socio-economic status. Since the traditional biomass energy such as crop straw and firewood is free of charge (Gosens et al., 2013), the poor households that cannot afford the high prices and costs of the advanced fuels are highly likely to rely on it for living. However, the traditional biomass fuels need to be collected and processed by households before use. That means the household, as an energy producer, has to allocate labor input to biomass collection and energy preparation. In this case, it could be restricted from engaging in other income-generating activities, given a fixed labor endowment (Van der Kroon et al., 2013). Thus, it is of great importance for a household to optimally allocate its labor on different activities to get returns for livelihood enhancement. Considering what has been described, we should simultaneously and holistically study the household biomass energy use behaviors from both production and consumption sides in this research.

In a growing body of literature focusing on the household behaviors towards biomass energy use, the theory of opportunity cost of labor were commonly used to examine the household decision-making behaviors between biomass collection and other activities in different developing countries. Amacher et al. (1996) used the micro data from Nepal to investigate the different responses of the households who collect firewood and participate in the firewood market to the changes in the opportunity costs of labor and found that the labor opportunity cost is one of the most important factors that determine the household behaviors. Mekonnen (1997, 1999) conducted an empirical study in Ethiopia to inspect the biomass fuel collection and consumption behaviors of the households using non-separable agricultural household model and emphasized that the labor time variable has significant impacts in all cases. Heltberg et al. (2000) and Mishra (2008) analyzed the household behaviors of collecting and using biomass energy such as agricultural residues and firewood in India. Their findings indicate that household choice of collecting biomass energy is based on shadow wage, which is determined by the opportunity cost of biomass collection. Fisher et al. (2005) examined the determinants of household activity choice, influencing forest use among the poor households in Malawi and revealed that the return from different activities can affect household choice behaviors. In the Chinese case studies, Chen et al. (2006) and Démurger and Fournier (2011) tested the household biomass energy consumption behaviors and also provided evidence to support that the shadow wage of household labor and the shadow prices of the fuels are the main factors affecting household biomass energy consumption. Although all of these studies have pointed out that the opportunity cost of labor (shadow wage of the labor) is the dominant influencing factor for household decision-making on biomass energy use, they ignored the linkages between biomass energy consumption and other exogenous markets such as labor market and commercial energy market. Therefore, in this study, we assume that the shocks of exogenous markets mainly come from the changes in prices, and we will empirically analyze the adjustment of household biomass energy use behaviors according to these price changes based on the agricultural household model, which was developed and elaborated in Section 2.4.

Another main assumption for this chapter is that the market for biomass energy is absent in our study region. Normally, households jointly make their decision on biomass energy consumption and biomass collection. In order to use biomass energy as a substitute for the more expensive fuels to meet its energy demand, the household allocates a certain part of its limited time endowment on collecting the biomass, which comes from the wastes of its own agricultural production. The main structure of this chapter is organized as follows: We will briefly introduce the data used in our study in Section 5.2 and test the separability for households surveyed in our field survey in Section 5.3. The analysis of household biomass energy use responses to the changes in the exogenous markets will be conducted in Section 5.4. Finally, we will conclude the main findings of this chapter in Section 5.5.

5.2 Descriptive analysis

5.2.1 Context of Household Energy Consumption

The data used in this chapter were also collected from our field survey conducted in 2013. Of the total 556 households, 524 households are participating in agricultural production, i.e. typical agricultural households. Table 5.1 shows the energy consumption status of our sampled households. Generally, the energy consumed by these households consists of biomass energy and commercial energy. Traditional biomass energy (including crops straw and firewood) and biogas are the main types of biomass energy used currently. Biomass energy takes the largest share (76.31% and 74.61%) of energy consumption of households from mountainous areas (Aba) and hilly areas (Yibin), while commercial energy occupies the dominant position (62.87%) in the household energy consumption of plain areas (Deyang). In particular, for biomass energy, the market is almost absent. In other words, households usually collect biomass and then prepare energy for residential use by themselves.

Table 5.1 Energy consumption status of sampled households (Kgsce per year per household)

	Aba		Yibin		Deyang	
	Amount	Share (%)	Amount	Share (%)	Amount	Share (%)
Traditional biomass energy	1775	73.53	1430	65.42	299	27.89
Biogas	67	2.78	201	9.19	99	9.24
Commercial energy	572	23.69	555	25.39	674	62.87

Source: Author's own field survey

With respect to household traditional biomass energy consumption, it can be seen from Table 5.2 that, on average, the households from mountainous areas consume the largest amount of firewood, while the households in plain areas use the least amount of firewood (339 kg per year). Due to the geographic and weather conditions, crops such as rice and wheat cannot be cultivated in mountainous areas. Thus, there are no enough available straws for the households to use as energy. However, the households in hilly areas have abundant crop straw resources. Therefore, they consume the largest amount of crop straw (442 kg per year).

Table 5.2 Traditional biomass energy cousumption of sampled households (Kg per year per household)

	Aba	Yibin	Deyang
Landscape	Mountainous areas	Hilly areas	Plain areas
Crops straw	4	442	217
Firewood	3105	2144	339

Source: Author's own field survey

Particulary considering the household consumption of biogas (see Figure 5.3), 312 of the agricultural households built biogas digestors in their houses, accounting for about 59.5% of the total, whereas only 239 households used biogas, occupying approximately 45.6%. The main feedstock for household biogas production is pig dung, and the biogas is mainly used for cooking. Moreover, as the annual average temperature of the mountainous areas is relatively low, it is not suitable for constructing biogas digestors there. Thus, both the number of household having biogas digestors and using biogas in mountainous areas are the smallest (89 and 53). In contrast, hilly areas (Yibin) have the largest number of biogas digestors as well as biogas users (131 and 106).

Table 5.3 Household use of biogas (Total sample size: 524 households)

	Aba	Yibin	Deyang
Number of households possessing biogas digestor	89	131	92
Number of households using biogas	53	106	80
Main using purpose	Cooking	Cooking	Cooking
Main resources for producing biogas	Pig dung	Pig dung	Pig dung

Source: Author's own field survey

Table 5.4 lists the commercial energy consumption status of the 524 sampled agricultural households. It can be found that households from Aba (mountainous areas) consumed the greatest amount of coal annually (1653 kg), while those living in plain areas used the least (9 Kg). However, households from plain areas consumed more of other types of commercial energy (LPG, natural gas and electricity) than those from the mountainous and hilly areas, because they are relatively wealthier and can afford the more expensive energy and costly energy use devices. In terms of electricity, Aba has the lowest average price of about 0.39 CNY per kWh, whereas Yibin has the highest level of 0.56 CNY per kWh. The main reason for this is that electricity in mountainous areas is mainly generated through small hydropower, while that in hilly areas depends on thermal power generation. More importantly, households from mountainous areas spend the least (943 CNY per year) on purchasing commercial energy, whereas those living in plain areas spend the most (1717 CNY per year).

Table 5.4 Commercial energy consumption of sampled households

	Aba	Yibin	Deyang
Consumption amount of commercial energy (per household)			
Coal (Kg per year)	165	29	9
LPG (Tank per year)	0.18	0.61	4.85
Natural gas (m^3 per year)	0	44	57
Electricity (kWh per year)	1712	1663	2018
Expenditure on commercial energy (CNY per year per household)			
Coal	284	26	12
LPG	31	82	582
Natural gas	0	83	108
Electricity	658	931	1050
Total expenditure on commercial energy (CNY per year)	943	1097	1717

Source: Author's own field survey

5.2.2 Varable Description

The socio-economic characteristics of the 524 households from three different regions are presented in Table 5.5. In terms of household head characteristics, households from Deyang in plain areas have the highest average age (54 years) and the highest educational level (7.01 years), whereas households located in mountainous areas (Aba) have the lowest average age (48 years) and the lowest educational level (5.89 years). The fraction of female household head in Aba is the highest (0.12), while in Yibin of the hilly areas, it is the lowest (0.03). With respect to the household demographic structure, households from Deyang have the smallest family size (3.7 persons); the largest fractions of male adults (0.45), female adults (0.46), and elderly people (0.18); and the smallest fraction of children (0.06). Households located in Aba have the largest family size (4.4 persons) and the smallest fractions of adult males (0.42) and elderly people (0.07), whereas households who are living in Yibin have the highest fraction of children (0.14) in their families.

Additionally, it can be clearly seen that households from Deyang have the highest income level (15305 CNY per year), the highest market wage rate (8.8 CNY per hour), and the lowest expenditure level (5712 CNY per year), whereas they possess the least areas of arable land (3.3 Mu). They also have the highest fraction of working members (0.87) and female working members (0.50) in their families. The average age (48.42 years) and schooling years (7.22 years) of their working members are the highest among three regions. Contrarily, households from Aba rely heavily on agriculture, as they have the greatest areas of arable land (8.87 Mu) and the highest returns from agricultural production (5407 CNY per capita) per year. Furthermore, they have the lowest market wage rate and the highest expenditure level (6815 CNY per year). The average fraction of working members in their families is the lowest (0.76), and they have working family members with the lowest average age (41) and the lowest average educational level (6.23). The households from Yibin have the smallest fraction of female working members (0.47). Moreover, they have the lowest per capita income level (13225 CNY per year) and the lowest income level of agricultural production (4216 CNY per year).

Table 5.5 Socioeconomic characteristics of sampled households

	Aba	Yibin	Deyang
Sample size	175	179	170
HH characteristics			
Fraction of females in HH	0.12	0.03	0.07
Average age	48	53	54
Average educational years	5.89	6.30	7.01
Household composition			
Household size	4.4	4.3	3.7
Fraction of adult males	0.42	0.44	0.45
Fraction of adult females	0.40	0.39	0.46
Fraction of children (≤14)	0.13	0.14	0.06
Fraction of elderly (≥65)	0.07	0.11	0.18
Household working members			
Fraction of working members	0.76	0.81	0.87
Fraction of female working members	0.49	0.47	0.50
Average educational years of working members	6.23	6.67	7.22
Average age of working members	41	47	48
Economic conditions of household (2012-2013)			
Total income per capita (CNY)	13400	13225	15305
Wage rate (CNY per Hour)	7.1	7.8	8.8
Agricultural income per capita (CNY)	5407	4216	4343
Total expenditure per capita (CNY)	6815	5967	5712
Land area (Mu)	8.1	5.9	3.3

Source: Author's own field survey

Finally, Table 5.6 shows the household time allocation for different activities. Households from the mountainous areas spend the longest time (841 hours) on agricultural production, while the households in plain areas spend the shortest time (667 hours) on farm work. Correspondingly, households in mountainous areas allocated more time (382 hours per year) on biomass collection than those from the other two areas, whereas households who are living in the plain spend less time (75 hours per year) collecting biomass. For off-farm employment, households from Yibin allocate the longest time (4551 hours per year) to off-farm work on average, while households from Deyang spend the shortest time (3716 hours per year).

Table 5.6 Household time allocation to different activities (hours per year)

	Aba	**Yibin**	**Deyang**
Time allocation on farm work	841	771	667
Time allocation on off-farm work	3727	4551	3716
Time allocation on biomass collection	382	261	75
Fraction of time allocation on biomass collection	0.10	0.07	0.04

Note: Commercial energy refers to coal, electricity, natural gas, LPG, and centralized supplying biogas.
Source: Author's own field survey

5.3 Separability

5.3.1 Model Specification

According to the theoretical model that has been discussed before, separability of production and consumption obviously brings different impacts on the behaviors of rural households compared with non-separability. However, the previous empirical studies focusing on biomass energy using behaviors usually make the assumptions that separable or non-separable properties hold. Due to the complexity of Sichuan's situation, the testing for separability should be done in this research. There have been two well-known global test methods. The first one is to estimate the profit or labor demand function and to find out whether the independent variables influence consumption decisions but not production decisions (Lopez, 1984; Benjamin, 1992; Bowlus and Sicular, 2003; Henning and Henningsen, 2007), and the other one is to estimate the marginal productivity (shadow wage) by estimating the production or cost function and to compare it with the market price (Jacoby, 1993; Skoufias, 1994; Le, 2010).

Considering the heterogeneity across households and the specific failure on one particular market, Carter and Yao (2002) and Dutilly-Diane et al. (2004) tested the reduced form of labor demand function with regime-specific participations based on the observed market participation behaviors of the households, while Bhattacharyya and Kumbhakar (1997) used the structural form of a production function to estimate the idiosyncratic shadow price with the unknown sample separation and without the specific failure to any market and compared them with observed effective market price for each household. Furthermore, Vakis et al. (2004) proposed a mixture model approach-based test at the reduced form level to assign the probability of being exposed to market failures (behaving under non-separation properties) to each household on the basis of unknown sample separation estimation.

For this research, due to the sampling procedure, the heterogeneous behaviors of households should not be ignored, and the global test for separability has limited usefulness. Compared with other idiosyncratic testing methods that recognize heterogeneity, the mixture model approach detects non-separability on all markets at once and avoids reliance on households' labor market participation to reach the conclusion of separability between the production and consumption decisions (Vakis et al., 2004). Therefore, a finite mixture model (FMM) is adopted to test the separability for the sample households. It provides a natural representation of heterogeneity in a finite number of latent classes. Considering the data availability and the data variation across the sample, we implement the test for separability from the perspective of agricultural production. Firstly, we set the null hypothesis of this study that a household will determine the labor allocated on farm work in accordance with one of two alternative regimes defined by expressions (2.18) and (2.35). Then, following Quandt and Ramsey (1978) and Greene (2012), the sample behavior can be characterized as a switching regression system with two component latent class models:

$$l_i^k = \begin{cases} l_i^1 = x'_{1i}\beta + \mu_{1i} \text{ if } \lambda_i^* < 0 \\ l_i^2 = x'_{2i}\gamma + \mu_{2i} \text{ if } \lambda_i^* \geq 0 \end{cases}$$

$$\lambda_i^* = x'_{\lambda i}\xi + \mu_{\lambda i} \tag{5.1}$$

Where $\mu_{1i} \sim N(0, \sigma_1^2)$, $\mu_{2i} \sim N(0, \sigma_2^2)$, $\mu_{\lambda i} \sim N(0, \sigma_\lambda^2)$ ($\sigma_\lambda = 1$) and ξ, β, γ, σ_1^2 and σ_2^2 are unknown. l_i^k is the observed response variable for each observation referring to the amount of hours allocated by households to on-farm work. l_i^1, l_i^2, and λ_j^* are latent unobserved variables. We still normalize all of the prices by agricultural output to simplify our analysis. Using the results of theoretical model, we have:

$$x_{1i} = \{w_i, B_i, Z_i\}$$
$$x_{2i} = \{w_i, B_i, Z_i, S_i, a_i\} \qquad (5.2)$$
$$x_{\lambda i} = \{w_i, B_i, Z_i, S_i, a_i\}$$

Where $x_{\lambda i}$ is the classification vector of variables set to help explain the class probability. Given that we cannot identify a priori the regime that a household participated in, a randomly selected observation l_i^k (household i's labor allocated to agricultural production) will have the probability of $\Pr ob(l_i^1 | x_{\lambda i}) = \Phi(x'_{\lambda i} \xi)$ belonging to the first regime and the probability of $\Pr ob(l_i^2 | x_{\lambda i}) = 1 - \Pr ob(l_i^1 | x_{\lambda i}) = \Phi(-x'_{\lambda i} \xi)$ belonging to the second one.

The mixture of normal distribution problem can be formulated as:

$$f(l_i^k | x_{\lambda i}) = \left(\frac{\Pr ob(l_i^1 | x_{\lambda i}) \exp[-\frac{1}{2}(l_i^1 - x_{1i}\beta)^2 / \sigma_1^2]}{\sigma_1 \sqrt{2\pi}} + \frac{[1 - \Pr ob(l_i^1 | x_{\lambda i})] \exp[-\frac{1}{2}(l_i^2 - x_{2i}\gamma)^2 / \sigma_2^2]}{\sigma_2 \sqrt{2\pi}} \right)$$

Therefore, the log-likelihood for our mixture of normal distributions is as follows:

$$\ln L = \sum_{i=1}^{n} \ln L_i = \sum_{i=1}^{n} \ln \left(\Phi(x'_{\lambda i} \xi) \frac{\exp[-\frac{1}{2}(l_i^1 - x'_{1i}\beta)^2 / \sigma_1^2]}{\sigma_1 \sqrt{2\pi}} + \Phi(-x'_{\lambda i} \xi) \frac{\exp[-\frac{1}{2}(l_i^w - x'_{2i}\gamma)^2 / \sigma_2^2]}{\sigma_2 \sqrt{2\pi}} \right)$$

Our problem now is to estimate the parameters $\{\beta, \gamma, \xi, \sigma_1, \sigma_2\}$ from the sample of N observations on $\{l_i^k, x_{1i}, x_{2i}, x_{\lambda i}\}$, $i = 1,\ldots, N$. Then, we can predict class membership by calculating the posterior probability:

$$\Pr ob(l_i^1 | x_{\lambda i}) = \frac{f(l_i^1 | x_{\lambda i})}{f(l_i^k)} = \frac{f(l_i^1 | x_{\lambda i}) \Pr ob(l_i^1 | x_{\lambda i})}{f(l_i^k | x_{\lambda i})} = \frac{\Phi(x'_{\lambda i} \xi) \frac{\exp[-\frac{1}{2}(l_i^1 - x'_{1i}\beta)^2 / \sigma_1^2]}{\sigma_1 \sqrt{2\pi}}}{L_i} \qquad (5.3)$$

For estimating the parameters of the FMM model, we use the ML (maximum livelihood) method assuming that the class probability is constant.

5.3.2 Data and Estimation

In order to estimate the above model, we firstly specify the two regimes according to the theoretical model in Section 2.4. Under the separation assumption, a household's decision on labor allocation depends only on production decision. In other words, the household characteristics and the factors that influence consumption should not affect it. It also implies that these variables could only impact the non-separable (constrained) regime. Then, for the classification vector, $x_{\lambda i} = \{w_i, B_i, Z_i, a_i\}$, we include the market wage rate of household labor (w_i), household characteristics affecting household consumption and preference (a_i), and production side characteristics (B_i, Z_i). We also include household location dummy variables (r_1, r_2) that can reflect the regional differences in our model.

Based on the FMM model, we apply the maximum likelihood method that has been discussed before to divide the whole sample households into two groups[15]. The dependent variable is the amount of hours spent on farm work. The descriptive information about the variables used in the test for separability is listed in Table 5.7.

Table 5.7 Descriptive information about variables in the FMM model

Variables	Description (Unit)		Mean	Std. Dev.
L_a	Hours spent on-farm (Hours/Year)	l_i^k	760.69	489.39
GHH	Household head gender (male=1)		0.93	0.26
AHH	Household head age (Years)		51.62	11.76
ASHH	Household head age squared		2806.73	1254.45
EHH	Household head education (Years)		6.39	3.46
FS	Family size (number)	a_i	4.13	1.36
FAM	Fraction of adult males		0.44	0.16
FAF	Fraction of adult females		0.41	0.15
FC	Fraction of children (≤14)		0.13	0.16
FE	Fraction of elderly (≥65)		0.12	0.23
MW	Market wage rate (CNY/Hour)	w_i	7.90	4.58
AL	Cultivated arable land areas (Mu)		4.13	3.71
P_{ci}	Price index of intermediate inputs (CNY/Unit)	B_i	7.23	13.81
OI	Non-labor income (CNY/Year)	a_i	2682.49	4428.59
DB	Distance to biomass collecting spots (Km)	Z_i	2.22	4.50
r1	=1, if the household is from mountainous areas		0.32	0.47
r2	=1, if the household is from plain areas		0.31	0.46
No. of Obs.	524			

Source: Author's own field survey

---

15 We define the non-separable regime as the constrained group whilst defining the separable regime as the unconstrained group.

Table 5.8 presents the results of the parameter estimation for the finite mixture model (FMM). The first column shows the results of the OLS estimation of the regression model on the whole sample. It can be seen that the agricultural production side characteristics such as arable land owned and the price index of intermediate inputs significantly influence the on-farm labor allocation. Moreover, the pooled OLS estimates also reveal the significant impacts of household characteristics (the fractions of elderly people, adult females, and adult males in the family) on the household decision of on-farm labor supply. The first column indicates that non-separability is an important issue for sample households. The second and third columns list the estimation results of the finite mixture model. According to the parameter estimates of FMM, it can be recognized that a household has a probability of 0.22 being categorized into the group titled component 1. This also means that 78% (409) of the sampling households belong to the other group, component 2. We can see that, for all households in both groups, the labor allocation decisions on on-farm work are not only affected by production-side characteristics but also by consumption-side ones. Then, the null hypothesis of separability on them is rejected. In other words, households in our sample behave in a non-separable manner.

Particularly, the estimates of the FMM model reveal that, for both components, the variables pertaining to agricultural production (the ownership of arable land and the price index of intermediate inputs) can significantly increase the labor allocated to farm work. The gender of the household head is an important influencing factor for most households. In terms of household characteristics, the demographic structure such as fractions of children and male adults and household location can significantly affect labor allocation on farm work for the surveyed households from component 2. In addition, for households from component 1, the fraction of adult female members significantly influences household time allocation to farm work.

Table 5.8 On-farm Labor Allocation: OLS and FMM estimates

Variable	OLS	FMM	
		Component1	Component2
Family size	7.5933	36.8665	−11.4057
Household head age	−6.4963	4.9613	1.7532
Household head age squared	0.0241	−0.1403	−0.0245
Household head gender	60.6066	18.8228	86.6190*
Household head education	−3.3194	−31.6284	4.2814
Fraction of elderly (≥65)	−152.9182*	−443.0192	−73.4927
Fraction of children (≤14)	519.4996	830.9659	333.8743***
Fraction of adult females	595.8271***	1454.129**	185.3537
Fraction of adult males	571.6494***	889.9482	290.9799**
Cultivated arable land areas	32.3617***	81.4009***	14.0322***
Market wage rate (log)	−6.4473	−13.34	7.8255
Distance to biomass colleting spots	−3.8346	−2.4395	−1.7233
Non-labor income (log)	30.3956**	87.9086	10.2386
Price index of intermediate inputs (log)	253.1725***	138.6342*	270.0458***
Households from mountainous areas (r_1=1)	102.9704**	−302.3485	257.3383***
Households from plan areas (r_2=1)	66.6860	−145.488	149.6454***
Constant	−329.5545	−951.8361	−278.4495
No. of Obs.	524	115	409
Sample proportion	1.0	0.22	0.78
R^2 (pooled OLS)	0.4153	−3721.8204	
Log likelihood (mixture)			

Note: The dependent variable of the model is total hours of the household spend on-farm (Unit: Hours). The missing dummy for regions is hilly area. The significance levels are defined as *10%, **5%, and ***1%.

5.4 Household Behavior Analysis

By running the FMM model, we have found that, in our sample household, consumption and production decisions are non-separable. In order to address the research question of how household biomass energy use reponses to changes in exogenous markets and what factors influence theses behavioral responses, we specify a household model that can be econometrically estimated. As the shadow wage cannot be directly observed, we have to adopt a two-stage modeling method proposed by many researchers (Lopez, 1984; Henning and Henningsen, 2007; Tiberti and Tiberti, 2012). Therefore, at the first stage, we estimate the shadow wage of household labor using the production function, and in the second stage, we include it in the household consumption system and labor share system to jointly analyze the households' behaviors.

5.4.1 Shadow Wage and Shadow Price Estimation

As it has been discussed in the theoretical section of this paper, under the conditions of non-separation, the shadow wage determines the household's decision on labor allocation. Therefore, we firstly estimate the shadow wage of the household's labor use production function using the same methods that have been adopted in Chapter 4 (See Section 4.3.1 and Section 4.4.1). The specification of the production system for household i is still defined as:

$$\ln TOA_i = \alpha_0 + \alpha_1 \ln TOB_i + \alpha_2 \ln L_{ai} + \sum \lambda_m \ln B_{mi} + \sum \gamma_k d_{ki} + \varepsilon_i \tag{5.4}$$

$$\ln TOB_i = \beta_0 + \beta_1 \ln TOA_i + \beta_2 \ln L_{bi} + \sum \theta_j d_{ji} + \mu_i \tag{5.5}$$

Where μ_i and ε_i are the error term.

In this model, TOA denotes the total value of agricultural products. TOB represents the total amount of the collected biomass. L_a is the total hours worked on farm. L_b is the time spent on biomass collection. B_m is a vector of other inputs for agricultural production, including the areas of arable land input (AL) and the total value of intermediate inputs (TCI). d_k and d_j are some other variables affecting agricultural production and biomass collection. Here, we include the distance to the nearest biomass collecting spots (DB) in biomass collection function. In addition, the location dummies (i.e. r_1 and r_2) are added into both equations.

The strategies for estimating the production system including the measures adopted to deal with the endogeneity problem and zero-value variables have been described in Chapter 4. We firstly use observable household characteristics such as age, gender, and educational level as proxies for management ability for both of the production activities. Secondly, we employ a bivariate probit model to estimate the household participation equations of biomass collection and off-farm employment. Based on the expression (2.37) derived from our theoretical model analysis, we assume that the household simultaneously and jointly makes participation decisions for biomass collection and off-farm work, and we specify the probit model as

$$Y_{in} = f(w_i, a_i, B_i, Z_i, E_i, r1, r2) \quad n = b, o \tag{5.6}$$

Where b represents biomass collection and O denotes working off-farm. Y_{ib} is 1 if the household collects biomass for energy use and 0 if the household does not. Y_{io} is 1 if the household engages in off-farm employment and 0 if it does not.

The variables of the bivariate probit are defined as before in the FMM model. Then the results of model estimation are used to compute the Inverse Mills Ratio (IMR) for each household i that participates in either of the activities:

$$IMR_{in} = \phi(w_i, a_i, B_i, Z_i, r1, r2) / \Phi(w_i, a_i, B_i, Z_i, r1, r2) \tag{5.7}$$

And for household i which does not participate in the activity, the IMRs are:

$$IMR_{in}^* = \phi(w_i, a_i, B_i, Z_i, r1, r2) / (1 - \Phi(w_i, a_i, B_i, Z_i, r1, r2)) \tag{5.8}$$

Finally, the parameter estimates of the production system are obtained by augmenting the regression with the IMR using IT3SLS. The descriptive information of the variables and data used in estimating the production system are presented in Table 5.9.

Table 5.9 Descriptive analysis of data in production system estimation

Variables	Description	Mean	Std. Dev.
TOA	Total value of agricultural outputs (CNY)	18259	47874.38
TOB	Total amount of collected biomass energy (kgsce)	3737	4505.28
L_a	Total hours worked on farm (hours)	761	489.39
L_b	Total hours worked on biomass collection (hours)	241	314.07
		Units	Share (%)
Y_{bi}	Participation in biomass collection	394	75
Y_{oi}	Participation in off-farm work	453	86
No. of obs.		524	

Source: Author's own field survey. Other explanatory variables are described in Table 5.3.

In our sample of 524 farm households, 394 (75%) households participate in biomass collection, while 453 (86%) households' members are working off-farm. Furthermore, the total value of agricultural outputs produced by per household is on average 18259 CNY per year, whereas the total amount of biomass energy produced per household is 3737 kgsce. Regarding labor allocation decision, the average time spent on farm work is 761 hours per year by one household, whilst the average time working on biomass collection is 241 hours.

The estimation results of the bivariate probit regression are reported in Table 5.10. It can be seen that, for biomass collection, the household location plays a vital role in determining household participation decision. Households located in plain areas are less likely to collect biomass than those from hilly areas. The longer the distance between a farmer's house and a biomass collecting spot is, the more likely the household will decide to collect biomass.

The possible reason for this phenomenon is that households located in remote mountainous areas usually collect firewood in the forest designated by the local government. In most cases, the forest is far away from their houses. In addition, the significant and positive coefficient of non-labor income implies that raising household non-labor income level increases the probability of participating in biomass collection.

Table 5.10 Estimates of Seemingly Unrelated Probit (SUP) model for biomass collection and market participation for sampled households

Variables	Participate in biomass collection (Y_{bi})	Participate in off-farm work (Y_{oi})
	Coef. (Std. Err.)	Coef. (Std. Err.)
Family size	0.0806 (0.0628)	0.5158 (0.829)***
Fraction of female adults	−0.6795 (0.7584)	1.8610 (0.8304)**
Fraction of male adults	0.2954 (0.7483)	1.9681 (0.8029)**
Fraction of children	0.4003 (0.7976)	0.2095 (0.7883)
Fraction of elderly people	0.5832 (0.4191)	−1.0591 (0.3572)***
Age of household head	0.0604 (0.0490)	−0.0076 (0.0481)
Age squared of household head	−0.0004 (0.0005)	0.0001 (0.0004)
Gender of household head	−0.4642 (0.3028)	0.2840 (0.2942)
Educational level of household head	−0.0311 (0.0231)	−0.0096 (0.0280)
Mountainous areas	0.1006 (0.2382)	−0.4088 (0.2360)*
Plain areas	−0.8809 (0.1664)***	−0.2365 (0.2013)
Areas of arable land owned	−0.0170 (0.0255)	−0.0065 (0.0240)
Distance to biomass collecting spot	0.5413 (0.1130)***	0.0438 (0.0378)
Market wage rate (log)	−0.1502 (0.1155)	0.0651 (0.0792)
Non-labor income (log)	0.1409 (0.0598)**	−0.1437 (0.0719)**
Price index of intermediate inputs (log)	0.0677 (0.0980)	−0.2252 (0.0946)**
_cons	−1.7461 (1.6193)	−0.9040 (1.5476)
No. of Obs.	524	524
Rho (1,2)	0.0260 (0.1277)	
Log pseudolikelihood	−221.50639	
Wald chi2 (32)	187.50***	

Note: Values in parentheses are standard errors, and the significance levels are *10%, **5%, and ***1%.

Regarding the participation decision on off-farm work, family size and the fraction of adult members have positive influences. Households with a larger fraction of adult members are more likely to find jobs in nearby cities, whereas households with a smaller fraction of elderly people are more likely to work off-farm. Moreover, the non-labor income level and the price of intermediate inputs are also important factors that affect household decisions on working off-farm. The increasing non-labor income could impel them to stay in their farms. Thus, the probability of working outside will decrease. Finally, if the price of the intermediate inputs increases, households will purchase more of them and then are less likely to leave their farms for work.

Table 5.11 Estimation results of the simultaneous equations using IT3SLS

Independent variables	ln_TOA	ln_TOB
Total value of agricultural outputs (log)		0.0478 (0.0796)
Total amount of biomass (log)	−0.0225 (0.1265)	
Hours working on agricultural production (log)	0.6073 (0.0933)***	
Hours working on biomass collection (log)		0.3545 (0.0465)***
Areas of arable land owned (log)	0.3046 (0.0813)***	
Total value of intermediate inputs (log)	0.0200 (0.0146)	
Age of household head	0.0393 (0.0289)	0.0074 (0.0303)
Age squared of household head	−0.0004 (0.0003)	−0.0001 (0.0003)
Gender of household head	0.0794 (0.1731)	0.1102 (0.1764)
Educational level of household head	0.0462 (0.0150)***	0.0197 (0.0153)
Distance to biomass collecting spot		0.0032 (0.0123)
Mountainous areas	0.4389 (0.1212)***	−0.1739 (0.1371)
Plain areas	0.4460 (0.1603)***	−0.5732 (0.1405)***
_cons	3.1859 (1.3669)**	5.8061 (0.9992)***
Sample selection (IMR$_h$)	0.4586 (0.2178)**	−0.5076 (0.2158)**
Endogenous variables[a]	ln_TOA, ln_TOB	
R^2	0.3698	0.2890
No. of Obs.	394	

Note: The missing dummy for regions is hilly areas. The significance levels are *10%, **5%, and ***1%. Values in parentheses are standard errors of estimated parameters. All other variables in this system are treated as exogenous to the system and uncorrelated with the disturbances. The exogenous variables are taken as instruments for the endogenous variables.

The IT3SLS estimates of the agriculture-energy production system are reported in Table 5.11. Most of the estimated parameters have the expected signs. With respect to agricultural production, the inputs including labor and arable land have significantly positive effects on agricultural output. Considering the characteristics of household head, the educational level has a significantly positive influence on farm production. Raising the educational level of the household head increases the agricultural outputs. Furthermore, the location dummy of mountainous areas significantly and positively affects agricultural production. Compared to the households from hilly areas, those located in mountainous areas have higher yields of agricultural products.

On the other hand, in biomass collection function, the labor input has a significantly positive effect on biomass collection. Also, the household allocation dummy of plain areas is significant. It indicates that households from plain areas are less likely to engage in biomass collection than those from hilly areas.

After getting the parameter estimation results of the production system, the shadow wage of the household labor and the shadow prices of collected biomass are collected using formulas (4.3) and (4.4) and are presented in Table 5.12.

5.4.2 Household Consumption Decisions

5.4.2.1 Model Specification

For the purpose of modeling household consumption decisions, we adopt the approach mainly following the framework of Beznoka (2014). The consumption decision system of each household can be specified by an almost ideal demand system (AIDS, see Deaton and Muellbauer, 1980) model.

Let ES_i denote the expenditure share of the ith goods, then the demand for consumption goods i is represented by the following system of equations (Buse, 1994):

$$ES_i = \alpha_i + \sum_j \gamma_{ij} \ln(p_j) + \beta_i \ln \frac{Y}{P^*} \tag{5.9}$$

Where Y indicates full income (for the households in constrained group, it refers to shadow full income), p_j denotes the consumer prices of goods j, and α_i is the good-specific constant. β_i is the parameter of the budget effect of demand, and γ_{ij} are the parameters of the effects of relative price changes. Then we assume that the translog price index ($\ln P^*$) can be approximated by a linear price index, i.e. by the Stone's price index ($\ln P^* = \sum_k ES_k \ln p_k$) suggested by Deaton and Muellbauer (1980), resulting in a linearized AIDS. However, this index could be seen as an endogenous variable, because it depends on a household's expenditure shares. Therefore, we replace the individual expenditure shares with the sample mean ($\overline{ES_k}$). We thus can estimate the linear approximation of the AIDS (LA/AIDS) as follows:

$$ES_i = \alpha_i + \sum_j \gamma_{ij} \ln(p_j) + \beta_i (\ln Y - \sum_k \overline{ES_k} \ln p_k) \tag{5.10}$$

The restrictions imposed on the AIDS model are in the following sets:

Adding-up: $\sum_i \alpha_i = 1 \quad \sum_i \beta_i = 0 \quad \sum_i \gamma_{ij} = 0$ (5.11)

Homogeneity in prices: $\sum_j \gamma_{ij} = 0$ (5.12)

Symmetry: $\gamma_{ij} = \gamma_{ji}$ (5.13)

Based on the estimation results of this LA/AIDS model, the compensated and uncompensated elasticites can be respectively calculated by using the formulas given in the work of Chalfant (1987) and Eales and Unnevehr (1988) as:

$$\sigma_{ij}^* = \sigma_{ij} + \overline{ES_j} + \hat{\beta}_i(\frac{\overline{ES_j}}{\overline{ES_i}}) = \delta_{ij} + \frac{\hat{\gamma}_{ij}}{\overline{ES_i}} + \overline{ES_j} \quad (5.14)$$

$$\sigma_{ij} = \delta_{ij} + \frac{\hat{\gamma}_{ij}}{\overline{ES_i}} - \hat{\beta}_i \frac{\overline{ES_j}}{\overline{ES_i}} \quad (5.15)$$

Where if $i = j$, $\delta_{ij} = 1$, otherwise $\delta_{ij} = 0$. The average expenditure shares of the households are denoted by $\overline{ES}$. The $\hat{\beta}_i$ and $\hat{\gamma}_{ij}$ are the estimated parameters in LA/AIDS model. Moreover, the formula used to calculate the full income elasticity for i-th good is (See Green and Alston, 1990):

$$\theta_i = \frac{\hat{\beta}_i}{\overline{ES_i}} + 1 \quad (5.16)$$

According to our theoretical model and the model specification for AL/AIDS discussed before, for conciseness and estimation reasons, we grouped household expenditure (consumed commodities) into 5 categories: domestically produced agricultural commodities (*a*, including rice, maize, wheat, rapeseed, vegetables, and livestock products), biomass energy commodities produced by household (*b*, composed of firewood and crop straw), commercial energy commodities purchased by household (*e*, consisting of electricity, coal, natural gas, LPG, and centralized supplied biogas), leisure time consumed by household (*l*), and other market commodities purchased by household (*o*).

5.4.2.2 Empirical Strategy

The LA/AIDS model is estimated by using the constrained iterative seemingly unrelated regression (Constrained ITSUR) method. This method allows the estimation of contemporaneous correlation in error terms across system equations, which then is used to derive more efficient estimates (Zellner, 1962; Zellner and

Huang, 1962; 1963). In the ITSUR procedure, an equation is excluded from the system of simultaneous systems. The parameters of the excluded equation can be identified in terms of the parameters of other equations using the add-up restriction, since the sum of the shares of expenditure is equal to 1. We drop out the demand equation for the group of other market commodities (ES_o). The restrictions of homogeneity in prices and symmetry are imposed on the model as constraints. However, some important issues concerning the demand system estimation must be considered. Firstly, the construction of the prices of the grouped commodities will be presented in the following section. Furthermore, the treatments of potential sample selection bias and endogeneity problems will be discussed.

- **Price of Grouped Commodity**

We collected the cross-sectional data of the households on the expenditures and prices of different commodity categories in our field survey. For households living in the same region, the variations in prices are quite small. Therefore, we have to set up household-specific prices by using sub-group consumption structure under the assumption that the expenditure shares of the commodities in the same group are constant (Castellón et al., 2012; Beznoska, 2014). Then, the prices are calculated by the sum of weighted prices of each term in that category (Lewbel, 1989; Suharno, 2010; Castellón et al., 2012; Beznoska, 2014). It is defined as:

$$p_j = \sum_{n=1}^{N} p_n ES_{nj} \quad \forall n = 1, 2, \cdots, N \tag{5.17}$$

Where es_{nj} is the expenditure share of commodity n in commodity group j and p_n is the price of commodity n. These prices are calculated for the commodity category a (self-consumed agricultural products), b (biomass energy), e (commercial energy) and o (other market goods).

We firstly unify the price units of self-consumed agricultural products to CNY per kg and unify the different energy price units to the standard coal equivalent price unit (CNY/kgsce) through dividing the energy prices by the conversion coefficients. Then we use the above formula (5.21) to obtain the weighted prices for these three commodity groups. Particularly, for other market goods, we adopt the method suggested by West and Parry (2009) using the price data collected in the field to calculate a price index for the composite market goods.[16] In addi-

---

16 The data was collected from the official website of Scihuan government: http://www.sc.gov.cn/10462/10464/10594/10601/2013/10/8/10279526.shtml and our field survey.

tion, the shadow wages of household labor and the shadow prices of collected biomass which have been estimated using the production system are included in the model as the shadow prices of leisure time and biomass energy,[17] respectively.

- **Potential Sample Selection Bias and Endogeneity Problem**

The zero-expenditure problem occurs when the household does not consume a certain group of commodities defined before. In our case, zero expenditure only exists in group b (biomass energy) due to the household's decision of nonparticipation in biomass collection. Estimating the LA/AIDS demand system only with the households who supply labor to collect biomass would cause sample selection bias. In addition, another problem in the demand system estimation arises, since a relevant share of households in the samples who work both for agriculture and biomass energy production does not engage in off-farm work. As an important component of shadow income (Y), the wages of these households cannot be observed. Then, estimating the system only with the households who participate in all three activities may induce biased results. In order to solve those potential sample selection bias problems, we firstly replace the missing data on wages with the regional mean and then adopt the method proposed by Heien and Wessells (1990) to estimate the model specified as follows:

$$ES_i = \alpha_i + \sum_j \gamma_{ij} \ln(p_j) + \beta_i \ln\frac{Y}{P^*} + \sum_n \theta_{in} a_n + \rho_b IMR_b + \rho_o IMR_o^* + \varepsilon_i \quad (5.18)$$

Where the terms IMR_b and IMR_o^* have been determined in Section 5.4.1 with formulas (5.7) and (5.8).

Another endogeneity problem is caused by the correlation between the budget (shadow income) and the allocation of consuming goods and leisure. If the shadow wage is included in the model, the allocation of the budget depends on the decision of leisure-work, and then the endogeneity problem of the term $\ln(Y/P^*)$ becomes serious (Beznoska, 2014). Moreover, the shadow price of biomass energy and the shadow wage of household labor are unobservable and calculated on the basis of estimating the production system. Therefore, they are both endogenous variables in our model. As suggested by Beznoska (2014), we use the non-labor income and household head characteristics (age, gender, and educational level) as instruments for all three endogenous variables. The instruments are assumed to be exogenous in our model and less correlated with intrahousehold labor allocation. Finally, the estimated model is similar to

---

17 We also use formula (5.17) to calculate the weighted price of biomass energy.

a Three-Stage-Least-Squares (3SLS) model. At the first stage, the endogenous variables are respectively regressed on the instruments, all the other exogenous variables, and the local dummies. This process can be expressed in the following equations:

$$X_m^* = \delta_m \ln OI + \sum_k \rho_{mk} CHH_k + \sum_n \sigma_{mn} X_n + v_m \qquad (5.19)$$

Where X_m^* are the endogenous variables; $\ln OI$ is the natural logarithm of non-labor income and transfers; CHH_k denotes household head characteristics including age, gender, and educational level of the household head; v_m are error terms; and X_n is the set of all exogenous variables in our model, which contains the exogenous market prices, household characteristics such as family size, fraction of adult male, fraction of adult female, fraction of children and fraction of elderly, and household location dummies. In the second stage, the censored Seemingly Unrelated Regression (Censored SURE) method is applied to estimate the LA/AIDS model (Heien and Wessells, 1990).

The variables used in estimating the LA/AIDS are described in Table 5.12. Leisure takes the largest share of household annual expenditure (82.5%), whereas commercial energy occupies the smallest (0.62%). The weighted prices of self-consumed agricultural products, commercial energy, and other marketed goods are 4.01 CNY/Kg, 3.32 VNC/Kgsce and 597.15 CNY/Unit, respectively. The average shadow income of the sampled households is 247255.7 CNY per year, and then the average value of the term $\ln(Y/P^*)$ is 9.7.

Table 5.12 Description of variables used in estimating AL/AIDS model

Variables	Description	Mean	Std. Dev.
ES_a	The expenditure share of self-consumed agricultural products	0.047	0.053
ES_b	The expenditure share of biomass energy	0.012	0.018
ES_e	The expenditure share of commercial energy	0.008	0.073
ES_l	The expenditure share of leisure	0.769	0.157
ES_o	The expenditure share of other marketed goods	0.164	0.134
Pa	The price index of self-consumed agricultural products (CNY/Kg)	4.01	3.27
Pb	The (shadow) price index of biomass energy (CNY/Kgsce)	0.92	1.47
Pe	The price index of commercial energy (CNY/Kgsce)	3.32	1.01

Variables	Description	Mean	Std. Dev.
Pl(w*)	The shadow price of leisure (shadow wage rate) (CNY/Hour)	9.47	3.42
Po	The price index of other marketed goods (CNY/Unit)	597.15	279.83
Y	Household shadow income (CNY/Year)	247255.7	120714.3
No. of Obs.		524	

Source: Author's own household survey

- **Estimation Results**

We now use the Hausman test to check the endogeneity of $\ln(Y/P^*)$, shadow price of biomass energy and shadow wage of household labor (Spencer and Berk, 1981). The statistic of the Hausman test in Table 5.13 implies that these terms cannot be treated as exogenous. Therefore, we use the three-stage estimation method that has been stated before. The estimates of the LA/AIDS demand system are also listed in Table 5.13. It can be seen from the estimated parameters of the LA/AIDS that, for biomass energy consumption, the market price of commercial energy is one of the main influencing factors. With an increase in commercial energy price, the share of the expenditure on biomass energy will also increase. In terms of household characteristics, the fraction of female adults in members and household location are found to be important in determining biomass energy consumption decisions. Households living in mountainous areas or with a larger fraction of female adults consume more biomass energy than others.

Table 5.13 Parameter estimation of LA/AIDS model using censored SURE

Variables		Dependent variables				
		ES_a	ES_b	ES_e	ES_l	ES_o
$\ln\_P_a$	γ_{ia}	0.0134***				
		(0.0027)				
$\ln\_P_b^*$	γ_{ib}	0.0039***	0.0199***			
		(0.0008)	(0.0028)			
$\ln\_P_e$	γ_{ie}	0.0024***	−0.0002	−0.0028***		
		(0.0004)	(0.0012)	(0.0011)		
$\ln\_p_l^*$	γ_{il}	−0.0515***	−0.0322	−0.0037	0.1911**	
		(0.0028)	(0.0045)	(0.0019)	(0.0211)	
$\ln\_P_o$	γ_{io}	0.0228***	0.0064***	0.0029***	−0.0780***	−0.0942
		(0.0032)	(0.0025)	(0.0011)	(0.0170)	

Variables		Dependent variables				
		ES_a	ES_b	ES_e	ES_l	ES_o
$[\ln\_(Y/P^*)]^a$	β_i	0.0521***	0.0034	0.0048***	−0.2658***	0.2056
		(0.0093)	(0.0174)	(0.0076)	(0.1186)	
Family size	θ_{in}	−0.0121***	−0.0022	−0.0042**	0.1434***	
		(0.0029)	(0.0047)	(0.0021)	(0.0321)	
Fraction of female adults	θ_{in}	0.0694***	0.0163*	0.0028	−0.2442***	
		(0.0195)	(0.0084)	(0.0033)	(0.0596)	
Fraction of male adults	θ_{in}	0.1051***	0.0084	0.0073**	−0.2011***	
		(0.0197)	(0.0088)	(0.0033)	(0.0614)	
Fraction of children	θ_{in}	0.0376**	0.0103	0.0035	−0.1057**	
		(0.0175)	(0.0076)	(0.0029)	(0.0537)	
Fraction of elderly	θ_{in}	−0.0380***	0.0041	−0.0026	0.0706**	
		(0.0093)	(0.0045)	(0.0020)	(0.0279)	
Mountain areas	θ_{in}	−0.0219***	0.0051**	−0.0059***	0.0127	
		(0.0042)	(0.0024)	(0.0009)	(0.0129)	
Plain areas	θ_{in}	0.0070	−0.0031	−0.0008	0.0372**	
		(0.0056)	(0.0024)	(0.0010)	(0.0165)	
IMR_b	λ_b	−0.0202***	−0.0085***	0.0017	0.0649***	
		(0.0072)	(0.0032)	(0.0014)	(0.0217)	
IMR^*_o	λ_o	−0.0283***	−0.0014	0.0018**	0.0230	
		(0.0054)	(0.0023)	(0.0009)	(0.0159)	
_cons	α_i	−0.6118**	0.0041	−0.0850	5.4964	−3.8037
		(0.3006)	(0.1630)	(0.0731)	(1.1105)	
Hausman test statistic			145.68***			
No. of Obs.			394			

Note: a. Endogenous variable. The missing dummy for regions is hilly area. The values in parentheses are standard errors, and the significance levels are *10%, **5%, and ***1%. Subscript a denotes self-consumed agricultural products, b represents biomass energy, e refers to commercial energy, l is leisure, and o denotes other marketed goods consumed by households.

Table 5.14 Price and income elasticity estimated by LA/AIDS model (mean value)

	Direct price elasticity			Income elasticity	
	C	UC			
σ_a	−0.747	−0.874	θ_a	1.691	
σ_b	−0.014	−0.038	θ_b	1.165	
σ_e	−1.194	−1.212	θ_e	1.362	
σ_l	−0.066	−0.431	θ_l	0.579	
σ_o	−1.103	−1.568	θ_o	1.792	
Cross price elasticity					
	C	UC		C	UC
σ_{ab}	0.073	0.038	σ_{bl}	−0.934	−1.761
σ_{ae}	0.045	0.022	σ_{bo}	0.570	0.268
σ_{al}	0.054	−1.121	σ_{el}	0.350	−0.509
σ_{ao}	0.563	0.124	σ_{eo}	0.478	0.125
σ_{be}	0.001	−0.014	σ_{lo}	0.136	−0.014
Cross price elasticity					
	C	UC		C	UC
σ_{ba}	0.265	−0.003	σ_{lb}	−0.031	−0.051
σ_{ea}	0.254	−0.116	σ_{ob}	0.045	0.024
σ_{la}	0.006	−0.088	σ_{le}	0.007	−0.006
σ_{oa}	0.163	0.073	σ_{oe}	0.024	0.011
σ_{eb}	0.002	−0.024	σ_{ol}	0.331	0.346

Note: C is compensated elasticity and UC is uncompensated elasticity. Subscript a denotes self-consumed agricultural products, b represents biomass energy, e refers to commercial energy, l is leisure, and o denotes other marketed goods consumed by households.

Equally important are the elasticities calculated on the basis of the estimation results of the LA/AIDS model (See Table 5.14). According to the demand theory, the compensated elasticity (Hicksian price elasticity) is derived from solving the dual problem of expenditure minimization at a certain utility level, assuming constant purchasing power, while the uncompensated elasticity (Marshallian price elasticity) is obtained from maximizing utility subject to the budget constraint. Both reveal the household's reaction on changes in prices of different commodities. In our case, we should pay more attention to the compensated elasticity, because our LA/AIDS model is deduced from the expenditure minimization problem.

With respect to the direct price elasticities of commodities groups, all the signs are negative. These results are consistent with the theoretical postulate. The expenditure on commodities will decrease when their prices increase. In terms of magnitude, self-consumed agricultural products, biomass energy, and leisure are price inelastic. That is to say, their expenditures are less responsive to the changes in their own prices. Regarding the compensated cross-price elasticities, the elasticities between biomass energy and self-consumed agricultural products (σ_{ab}) and commercial energy (σ_{be}) and other marketed goods (σ_{bo}) are positive, showing that the relationships between biomass energy and these three types of commodities are substitutes. Nonetheless, the cross-price elasticities between biomass energy and and leisure time (σ_{bl}) is negative, implying that biomass energy and leisure time are complements. Additionally, all other compensated cross-price elasticities are positive, which indicates that the relationship among the four different commodity groups (self-consumed agricultural products, commercial energy, leisure, and other marketed goods) should be substituted.

Furthermore, the positive signs of income elasticities demonstrate that all five types of commodities are normal goods. Among which, only the consumption of leisure time is less responsive to the changes in household shadow income level, and the expenditures on other commodities are relatively sensitive to household shadow income level. In particular, when household shadow income level increases, the biomass energy consumption level increases.

5.4.3 Household Labor Supply

5.4.3.1 Model Specification

On the other hand, in order to investigate the responses of households' labor supply to the changes in shadow wage rate (p_l^*) and market wage rate (MW), we use factor cost share equations, which are derived from a translog cost function to estimate the elasticity (Schneider, 2011). Following Fisher et al. (2005), we assume all income generation (including shadow income) activities as a process of production. Then, in the production function, we assume that there are three different inputs, labor working off-farm, labor allocated to domestic production (agricultural production and biomass collection), and the intermediate inputs for agricultural production (fertilizer, pesticides, and plastic films). According to Shephard's lemma, the demand for a factor with minimizing cost can be obtained from the differentiation of the cost function with respect to its price (Mascolell et al., 1995). Therefore, in our case, the system of a translog cost function with cost share equations is established in the following:

$$\ln TC = \alpha_0 + \alpha_i \ln p_i + \frac{1}{2}\sum_i \sum_j \gamma_{ij} \ln p_i \ln p_j + \alpha_y \ln Y_p + \mu_i$$

$$LS_j = \alpha_j + \sum_j \gamma_{jk} \ln p_j + \varepsilon_j \tag{5.20}$$

Where LS_j is the cost share of factor j and p_j is its price. α_j is the constant term and ε_j is the error term. Similar to the AIDS model, some restrictions have to be imposed on the system of equations as follows:

Adding-up restriction: $\sum \alpha_j = 1$ (5.21)

Homogeneity in prices: $\sum_j \gamma_{jk} = 0$ (5.22)

Symmetry: $\gamma_{jk} = \gamma_{kj}$ (5.23)

Additionally, to solve the problem of singularity, we drop the cost share equation of intermediate inputs from the system. Finally, a system of two translog equations will be directly estimated:

$$LS_n = \alpha_n + \gamma_{n1} \ln(\frac{w^*}{P_{ci}}) + \gamma_{n2} \ln(\frac{w}{P_{ci}}) + \varepsilon_n \quad n = 1, 2 \tag{5.24}$$

Where LS_1 is the cost share of labor allocated to domestic production and LS_2 denotes the cost share of labor engaging in off-farm employment. P_{ci} is the price of intermediate inputs. The subscript n represents the different regions for the local dummies (mountainous areas and plain areas). AL denotes arable land ownership. RDL is the ratio of the dependence to labors (in our sample, the mean value of RDL is 0.3217), and a_i denotes household characteristics including the demographic structure characteristics such as family size, fractions of male adults, female adults, children, and elderly people and household location dummies r_n (n = 1,2) representing mountainous areas and plain areas. In addition, in order to avoid the potential problems caused by sample selection bias, the IMRs calculated using the standard Heckman sample selection model (described in Section 4.4.1) are also included in both of the equations.

After obtaining the estimated coefficients of the system of equations (5.28) using the constrained iterative seemingly unrelated regression estimation (constrained ITSURE) method, we can calculate the own elasticity and substitution elasticity of labor, employing the formula given by Binswanger (1974):

$$E_{ii} = \frac{\gamma_{ii}}{LS_i} - 1 + \overline{LS_i} \tag{5.25}$$

$$E_{ij} = \frac{\gamma_{ij}}{LS_i} + \overline{LS_j} \qquad (5.26)$$

5.4.3.2 Estimation Results

As the unobservable shadow wage is endogenously determined within household, the estimation of the labor cost share equations with shadow wage is similar to the estimating process of instrumental variable regression adopted in estimating the SNQ profit function and the LA/AIDS model before. At this time, we choose the logarithm of non-labor income (ln OI) and the distance to the nearest biomass collecting spot (DB) as the instrumental variable. Firstly, we regress the term $\ln(w^*/p_{ci})$ on the instrumental variables and all the other exogenous variables. Then the predicted value of this term will be used as the augmented variable in the constrained IT3SLS in the second step. Table 5.15 lists the estimation results of the labor cost share equation system with the instrumental variables. Most of the coefficients are significant at the 1% level. IMR_a^* is significant in both labor share equations, indicating that the sample selection bias would happen if the system was estimated while taking into consideration households that do not participate in off-farm work. The positive sign of the market wage rate in the domestic production equation demonstrates that the households that can earn more wage income allocate less time to domestic production, while the negative sign of the shadow wage in the same equation reveals that households will supply more labor to domestic production activities if they can obtain higher returns from them. In contrast, in the off-farm employment equation, the household time allocation to off-farm work is negative to the market wage rate and positive to the shadow wage. Furthermore, we could also find that the household characteristics such as family size, fractions of adults and elderly people among household members, areas of owned areable land, and household location have significant impacts on household labor allocation. Households with larger family size and larger fractions of adults tend to spend more time on domestic production, while allocating less labor to off-farm employment. In contrast, households with larger fractions of elderly people spend more time working domestically and less time taking a part in off-farm employment. In addition, households possessing more arable land will increase the time allocated to domestic production and reduce the time allocated to off-farm work. Households located in mountainous areas supply more labor to domestic production and less labor to off-farm work than those from hilly areas.

Table 5.15 Constrained IT3SLS estimation results of labor share equations

Variables	Total Cost	Share of domestic (LS_1)	Share of off-farm (LS_2)
$\ln p_c$	0.164 (0.040)***		
$\ln p_f$	0.632 (0.048)***		
$\ln p_d^*$	0.204 (0.059)***		
$(\ln p_c)^2$	0.013 (0.023)		
$(\ln p_f)^2$	0.197 (0.031)***		
$(\ln p_d^*)^2$	0.069 (0.040)*		
$\ln p_c * \ln p_f$	−0.070 (0.029)**		
$\ln p_c * \ln p_d^*$	0.057 (0.036)		
$\ln p_f * \ln p_d^*$	−0.126 (0.014)***		
$\ln Y^*$	0.216 (0.131)*		
$\ln(p_f/p_c)$		−0.126 (0.014)***	0.192 (0.017)***
$\ln(p_d^*/p_c)$		0.099 (0.019)***	−0.126 (0.014)***
Family size	0.263 (0.071)***	−0.087 (0.020)***	0.052 (0.021)**
Fraction of adult females	0.679 (0.348)*	−0.025 (0.132)	0.138 (0.146)
Fraction of adult males	0.599 (0.354)*	0.063 (0.134)**	0.079 (0.149)
Fraction of children (≤14)	0.033 (0.561)	−0.039 (0.211)	0.214 (0.231)
Fraction of elderly (≥65)	−0.647 (0.344)*	0.113 (0.126)*	−0.023 (0.139)*
Areas of arable land owned	−0.001 (0.008)	0.008 (0.003)***	−0.004 (0.003)
Ratio of dependence to labor	−0.176 (0.124)	0.183 (0.102)*	−0.231 (0.111)**
Mountainous area ($r_1=1$)	0.087 (0.068)	0.076 (0.025)***	−0.145 (0.027)***
Plain area ($r_2=1$)	0.006 (0.078)	−0.023 (0.027)	−0.048 (0.029)*
IMR_o'	−0.150 (0.159)	0.036 (0.044)*	0.083 (0.048)*
_cons	4.581 (1.483)***	0.419 (0.118)***	0.235 (0.130)*
R^2	0.5337	0.3226	0.4053
Hausman test statistic		92.96***	
No. of Obs.		524	

Note: The missing dummy for regions is hilly area. The values in parentheses are standard errors, and the significance levels are *10%, **5%, and ***1%. The letter a indicates the endogenous variable. In our sample, the labor cost shares of domestic production and off-farm work are 0.284 and 0.561, respectively.

Based on the above results, we can compute the elasticities of household labor supply using the estimated parameters of the labor share equations. The mean value of the own price elasticities of labor supply to domestic production and off-farm employment are −0.366 (E_{11}) and −0.098 (E_{22}), respectively. As outlined by

Dogan (2008), the convexity of the translog cost function in factor prices based on the assumption of cost minimization requires the own-price elasticities of inputs to be negative. Therefore, the condition is satisfied in our estimation results. The cross-price elasticities of these two behaviors are 0.108 (E_{12}) and 0.054 (E_{21}), supporting the theoretical expectations on the signs of the production elasticities. These elasiticities also imply that the labor allocated to biomass collection will decrease when the shadow wage of the household rises, and if the market wage rate goes up, the household will reduce the time spent on biomass collection.

5.5 Household Biomass Energy Use Responses to the Changes in Exogenous Markets

As the behaviors of the households in our sample satisfy the properties of non-separable assumptions, the effects of the changes in exogenous markets on household biomass energy use are examined, based on the expressions of elasticities derived from the theoretical framework in Chapter 2:

$$E(w_i^* / p_{xi}) = -\frac{E_{xi}^* + \gamma_i(\sigma_{xi}^* + \theta_i^* S_{xi})}{E_i^* + \gamma_i(\sigma_i^* + \theta_i^* S_{li})} \quad (5.27)$$

$$E(C_{bi}/p_{xi})^G = E(C_{bi}/p_{xi})^H + E(w_i^*/p_{xi})[E(C_{bi}/w_i^*)^H + \theta_{bi} S_{li}] \quad (5.28)$$

The indirect effect occurs through changes of shadow wage in response to the changes of prices in the exogenous market (5.27), while the direct effect comes from the impacts of exogenous prices on biomass energy consumption (the first term in equation (5.28)).

With respect to the elasticity in (5.27), which reflects the changes of household shadow wage in response to the changes in prices of the commodities in exogenous markets, we can identify its sign at mean value based on the estimation results of our models. For the term in its denominator $E_i^* + \gamma_i(\sigma_i^* + \theta_i^* S_{li})$, the value of the compensated own-price elasticity of leisure consumption (σ_i^*) is −0.066, and the full income elasticity of leisure consumption (θ_i^*) is 0.579. The share parameter of leisure consumption in the shadow income of our sampled households is on average 0.699, while the average value of the ratio (γ_i) is about 29.995. According to the estimates of the labor cost share equations, the value own-price elasticity of labor supplied to domestic production (E_i^*) is −0.366. Thus, the sign of the denominator of elasticity (5.27) is unambiguously positive.

Turning to the term in its numerator ($E_{xi}^* + \gamma_i(\sigma_{xi}^* + \theta_i^* S_{xi})$), the compensated cross-price elasticities of the leisure consumption (σ_{xi}^*) with respect to all exogenous market prices are positive according to our LA/AIDS model. Therefore, the second term ($\gamma_i(\sigma^*_{xi} + \theta^*_i S_{xi})$) is unambiguously positive. For the cross-price elasticities of labor allocated to domestic production (E^*_{pi}), we can deduce their signs from the estimated compensated price elasticities of consumption due to the fact that production limits consumption due to absent markets. Firstly, the cross-price elasticities of biomass energy and self-consumed agricultural products with respect to the price of commercial energy and other marketed goods are positive, implying that, when the exogenous price of commercial energy (p_e) or other marketed goods (p_o) increases, the demand of biomass energy and self-consumed agricultural products will increase as well. Thus, households will spend more time on domestic production. In these cases, the numerator is positive and the expression of (5.27) ($E(w^*_i/p_{ei})$ and $E(w^*_i/p_{oi})$) is unambiguously negative.

Additionally, if the exogenous market price refers to the price of self-consumed agricultural products (p_a), the situation could be more complicated. Despite the fact that the signs of the compensated own-price elasticity of self-consumed agricultural products and the compensated cross-price elasticity of biomass energy to self-consumed agricultural products are opposite, $|\sigma_a| > |\sigma_{ba}|$ indicates that the term in the numerator should be negative. Thus, $E(w^*_i/p_{ai})$ is unambiguously positive. On the other hand, for the changes in exogenous labor market, due to the binding constraint imposed on off-farm employment, an increase in market wage rate results in an increase in labor allocated to home production (0.108) and a decrease in leisure consumption. Therefore, we can say the sign of the numerator in (5.27) is unambiguously negative, with the result that $E(w^*_i/w_i)$ is positive. In order words, the shadow wage of household labor should respond positively to the changes in the exogenous price of self-consumed agricultural products and labor, while responding negatively to the changes in other exogenous prices. Increasing the price of self-consumed agricultural products and labor increases the shadow wage, while increasing the price of commercial energy and other marketed goods can decrease the shadow wage rate.

Considering the elasticity in (5.28), the first term of its right-hand side (RHS) represents the direct effects of the changes in exogenous market on biomass energy use. Its sign is positive for all categories of goods in accordance with the estimates of the LA/AIDS model. The second term of the right-hand side (RHS) of elasticity (5.28) reveals the indirect influence of the changes in exogenous market on biomass energy use via internal adjustments. Since the term $\left[E(C^*_{bi}/w^*_i)^H + \theta^*_{bi}S_{li}\right]$ is unambiguously positive, the product of it and depends on the sign of. $E(w^*_i/p_{xi})$ depends on the sign of $E(w^*_i/p_{xi})$.

For the changes in market prices of self-consumed agricultural products and labor, $E(C_{bi}^*/p_{xi})^G$ is positive, as both terms in equation (5.28) are positive. On the contrary, for the household biomass energy use responses to changes in market price of commercial energy and other marketed goods, due to the opposed signs of the two terms on the RHS of the elasticity (5.28), its sign depends on the magnitudes of $E(C_{bi}^*/p_{xi})^H$ and $E(w_i^*/p_{xi})\left[E(C_{bi}^*/w_i^*)^H + \theta_{bi}^* S_{li}\right]$. As the magnitudes of $E(w_i^*/p_{xi})$ are ambiguous, the signs of $E(C_{bi}^*/p_{xi})^G$ are also ambiguous.

According to what we have discussed before, we can determine that, with an increase in the market prices of labor and self-consumed agricultural products, household biomass energy consumption will increase. In addition, in the case of the changes in the prices of commercial energy and other marketed goods, the sum of both direct and indirect effects $E(C_{bi}^*/p_{xi})^G$ is smaller than the direct effect itself if the indirect effect is negative. This indicates that the market failure reduces the flexibility of household behaviors.

5.6 Conclusions

In this chapter, we investigated household biomass energy use responses to the changes in exogenous markets. We firstly conducted a test for separability on the sampled households to examine whether their behaviors satisfy the property of separable assumptions. Based on this, we estimated the shadow wage of their labor and systematically examined the biomass production and consumption behaviors by jointly estimating the system of household labor share equations and the LA/AIDS model. The main findings obtained from the estimation results of these models are as follows:

(i) Households make their decisions on on-farm labor allocation according to the assumptions of non-separability. Specifically, among different influencing factors, areas of owned arable land and average price of intermediate inputs could significantly and positively affect the household labor allocation to agricultural production.

(ii) We have examined the factors influencing the household decisions on participation in biomass collection and off-farm work. The household characteristics such as family size; fraction of male adults, female adults, and elderly people; and household location have a significant influence on household participation in off-farm work. Households with older and less educated heads are more likely to use traditional biomass energy. Households who are living farther from the biomass collecting spots have a higher likelihood of engaging in biomass collection. Obviously, households located in mountainous areas are less likely to work off-farm than those who are from hilly areas, while households located in plain areas are less likely to collect biomass. The main reasons for this situation include the weather and geographical conditions, living customs, and the different development levels of the regional economies. The non-labor income level was found to have opposite effects on biomass collection and off-farm work. It is positive to biomass collection and negative to off-farm employment.

(iii) Our analysis indicates that biomass collection competes with agricultural production as it reduces the labor input for agricultural production. In reverse, the agricultural production could provide more biomass resources for household energy use. In this case, agricultural production could bring a positive effect on biomass collection. These findings are also in line with those found in Chapter 4.

(iv) The relationship between off-farm work and domestic production (including biomass collection) is also competitive. We also found that households with larger family size, a higher fraction of female adults, and a higher fraction of male adults have larger labor share in off-farm working activities and a smaller labor share in domestic production labor. Conversely, households possessing a larger fraction of elderly people and more arable land have a smaller labor share in off-farm employment and a larger labor share in domestic production. Additionally, compared to those from hilly areas, households from mountainous areas allocate more labor to off-farm work and less labor to domestic production activities.

(v) With respect to household biomass energy consumption decision-making behaviors, the shadow price of biomass energy and the shadow wage of household labors are two significant influencing factors. Concretely, an increase in shadow wage will reduce the expenditure on biomass energy, while an increase in the shadow price will increase the financial burden of the households to use biomass energy. The changes in the price of the commercial energy will bring a positive impact on the biomass energy expenditure, and when the income levels of the households increase, they will spend more on biomass energy. The household characteristics such as the fraction of adults, the fraction of children, and the fraction of elderly positively affect the expenditure on biomass energy. The households from mountainous areas spend more on traditional biomass energy than those from plain areas.

(vi) Regarding the elasticities calculated from the results of the AL/AIDS model, we find that biomass energy is a kind of normal good with negative own price elasticity and positive income elasticity. However, it is price inelastic and is less responsive to the changes in its shadow price. The relationship between biomass energy and commercial energy is substituted. Additionally, with an increase in market prices of labor and self-consumed agricultural products, household biomass energy consumption increases, under the imperfect labor market.

Chapter 6　Conclusions and Policy Implications

6.1 Conclusions

Because biomass energy functions as a link between nature and human activities, it is a crucial element in the frame of rural livelihoods. We have determined that the energy transition from traditional biomass energy such as crops straw and firewood to modern energy sources at household level could improve livelihoods for rural households in terms of health, environmental protection, and income generation, etc. In other words, the factors and measures that decrease household use of traditional biomass energy can also positively affect livelihood. Therefore, with particular concern for traditional biomass energy, we evaluated the impacts of household biomass energy use on rural livelihoods.

As presented in Chapter 2, we established a theoretical framework based on an agricultural household model. This model provides a basic roadmap for us to analyze the decisions pertaining to energy use, agricultural production, biomass collection participation, biomass energy consumption, and the factors affecting household biomass use. Firstly, to clarify how households make choices regarding energy use for cooking, we treated households as consumers in the energy market and set a basic hypothesis that, with the improvement in households' socioeconomic status, their energy use choices will shift from solid traditional biomass energy to more advanced fuels. Then, examining the influence of biomass collection on agricultural production, we have emphasized that households could be considered price takers and profit maximize in the domestic production system, and we developed our hypothesis that the behavioral relationship between agricultural production and biomass collection is competitive. In the analysis of household biomass energy use responses to the shocks in the consumer market, we identified that households play a double role as supplier and consumer in biomass energy utilization and discussed household biomass energy use behaviors through the interlinks between consumption, production, and labor market participation under both separable and non-separable assumptions. Under this circumstance, we have proposed two hypotheses: for non-separable households, biomass energy use is determined by the shadow wage of household labor. As the shadow wage increases, the use of biomass energy increases as well. For separable households, biomass energy consumption is influenced by household income level and by the shadow price of biomass energy according to the consumer theory of microeconomics.

On the basis of our theoretical framework and basic hypotheses, we empirically provided a comprehensive quantitative analysis using data collected from a field survey in rural Sichuan. The main findings are discussed from Section 6.1.1 to Section 6.1.3.

6.1.1 Biomass Energy Choice in Energy Transition and Its Determinants

The current energy use pattern for cooking in our study region shows that a majority (95%) of the 556 sampled households use more than one type of fuel at the same time. Despite the fact that 528 (91.7%) households use electricity for cooking energy, biomass is a sort of commonly used energy source. 432 (77.7%) of the households still burn traditional solid biomass (crops straw and firewood), whereas 243 (43.7%) of them produce biogas using pig dung. This not only suggests that household energy transition from traditional biomass energy to advanced energy is still slow, but also provides evidence of fuel stacking, as households do not abandon the traditional solid biomass energy (crops straw and firewood) when they use other types of more advanced fuels.

In order to further understand how a household makes decisions when choosing a specific energy source for cooking, we collected data of households' actual energy choices in our field survey, conducted a discrete choice experiment to observe their stated choices, and then examined the determinants of household choice behaviors, jointly considering the revealed and stated preferences. The statistic information of the actual fuel choices made by our sampled households in Sichuan Province reveals that traditional solid biomass energy (crop straw and firewood) is preferred for cooking. The different energy choice behaviors of households from different income-level groups confirm that income is still an important determinant of cooking fuel choice. In addition, according to our empirical analysis, households prefer to adopt the fuels with lower cost, higher safety, and lower indoor air pollution. The characteristics of the decision maker such as age, education level, and marital status; the household demographic structure including the fractions of adults, children, and elderly people; and household location are the main factors affecting household energy choice behaviors.

6.1.2 Impacts of Biomass Collection on Agricultural Production

According to the real situation in our study region, we modeled household behavioral relationship between agricultural production and biomass collection by employing a multioutput production system. We found that, for biomass collection participation, the education level of the household head, the non-labor income level, and household location are the main important influencing factors. Our findings also demonstrate that household decisions regarding participating in biomass collection and agricultural production are interdependent. The possible reason for this could be that household members who work on-farm are also mainly responsible for collecting biomass. This could also be an underlying reason behind the relationship revealed by the estimation results of the system of production functions. The parameters of the production system indicate that biomass collection negatively influences agricultural production, whereas agricultural production could have a positive impact on biomass collection.

Furthermore, in the SNQ profit function, the supply (agricultural products and biomass energy) cross-price elasticities are negative, indicating a competitive relationship between these two activities. Meanwhile, the cross-price elasticities of the two inputs (labor and intermediate inputs) are also negative, representing that inputs such as fertilizers and pesticides are substitutes for labor inputs in Sichuan Province.

6.1.3 Biomass Energy Use Responses of Households to the Exogenous Market

We provided a holistic and comprehensive analysis of household biomass energy use behaviors in this research. Based on the theoretical framework of an agricultural household model, we tested separability using our sampled households. All of the households in our sample were found to behave in a non-separable manner. The potential impacts of the changes in exogenous consumer market on household biomass energy use are complex because household decisions on biomass consumption, biomass collection, and labor market participation are usually interlinked and interdependent. The markets for biomass energy and labor are missing or imperfect. The effects we investigated include the direct effects of exogenous shocks (i.e. changes in prices of commodities such as agricultural products, commercial energy, and other marketed goods) on labor supply and biomass energy demand reaction and the indirect effects on labor allocation and biomass energy consumption adjustment through the changes in the shadow wage of household labor. For the direct effects, the positive cross-price elasticities of biomass energy consumption with respect to the exogenous prices calculated on the basis of the estimates of our LA/AIDS model reflect that increasing the exogenous prices directly increases biomass energy use. Particularly in the case of changes in the exogenous labor market, the positive labor demand cross-price elasticity obtained from estimating the labor share equations reveals that an increase in market wage rate will increase household labor allocated to domestic production activities (agricultural production and biomass collection). Thus, in turn, such an increase may also increase the use of biomass energy. Regarding such indirect effects, the elasticities of the shadow wage to the price of commercial energy and other marketed goods ($E(w_i^* / p_{xi})$) are negative. Since the biomass energy consumption elasticity to shadow wage ($E(C_{bi}^* / w_i^*)$) is positive, therefore, the indirect effects of exogenous shocks from labor market and agricultural product market on biomass energy use are positive, while those of shocks from commercial energy market and other exogenous markets are negative. Then, as the total effects equal the sum of the direct and indirect effects, the response of household biomass energy use to the changes in prices of labor and self-consumed agricultural products are unambiguously positive, implying that, when these prices increase, household biomass energy use increases. Turning to the changes in price of commercial energy and other marketed goods, the total effects are smaller than the direct effect, indicating that the imperfect labor market reduces the flexibility of household behaviors.

Additionally, our findings also highlighted the important role of education and the dominating place of income level in biomass energy consumption in rural Sichuan Province. Moreover, the impacts of household location on production and consumption decisions indicate that regional effects should not be neglected.

6.2 Policy Implications

As discussed in Section 1.3.3, the existing energy policies for developing biomass energy implemented in Sichuan Province are mainly focused on financial support for biogas construction in rural areas. For the target of Sichuan local government to simultaneously decrease the traditional use of biomass energy and promote modern bioenergy technologies (such as biogas production, comprehensive use of crop straw, and biomass power generation), the energy polices need to be improved. In order to benefit household livelihoods by promoting energy transition while enhancing agricultural production, this study proposes the following suggestions for future rural energy construction in rural Sichuan.

6.2.1 Adjust Energy Price and Improve Energy Quality

On one hand, the government should not only provide subsidies on constructing a household biogas digestor, but should also use some other effective market instruments to adjust the prices of specific fuels. Concretely, the government should adjust the prices of some modern commercial energy, such as electricity and natural gas. Alternatively, the subsidies on the prices of clean and efficient fuels (including the relevant technologies) should be provided for promoting their use in rural areas. Decreasing the prices of the substituted energy sources would decrease the use of traditional biomass energy. Meanwhile, reducing the cost of new biotechnology adoption could increase the use of modern biomass energy.

On the other hand, the future energy policies should attach more importance to the combined effect on both price and quality of modern fuels rather than merely paying attention to either of them. Precisely, any sustainable energy policy should provide more incentives that reduce energy price (or energy usage cost) while improving energy quality at the same time. For example, the most effective way to promote biogas in Sichuan Province is to upgrade modern biotechnologies to shorten the time spent on operating and cleaning the digester and to improve the safety of the users.

6.2.2 Enhancing Households' Access to Modern Fuels

Poverty is a main factor restricting households from obtaining modern fuels. In order to help poor households escape poverty and adjust the rural energy consumption structure, any strategies that provide alternative livelihood options for them would enhance their access to modern fuels. In the long term, the government should invest more in rural education. More skill trainings related to the operation and maintenance of the biogas digester or other modern energy devices should be provided. More information about the upgrading of latest technologies should be provided to households. Moreover, the government should formulate policies to create more job opportunities for rural households aiming at increasing their income levels. This would lead to more use of modern commercial energy.

6.2.3 Eliminating the Market Faliures

Since biomass energy plays an important role in rural livelihoods in China, the elasticity of household biomass energy use response to price incentives is crucial for food security enhancement and economic development. Thus, the results we have obtained indicate several elements of policy interventions that can be used to control this elasticity. One is the role of measures directed at reducing the incidence of market failures for specific households. This includes interventions that have capability to mitigate the binding constraint imposed on off-farm employment such as increasing investments in infrastructure construction, promoting education in rural areas and smoothing circulation of information on wages and job opportunities. Moreover, indirect sources of market failure also need to be eliminated by establishing a sound and effective social safety net to provide better access for rural households to services such as public transport system, health care, landless employment guarantee.

6.2.4 Increasing Attention to Regional Differences

As household location is a key factor affecting household use of biomass energy, the regional difference should be taken into account when designing new policies. Different regions have different situations. Therefore, the proposed energy policies must be adjusted to local conditions. In the context of Sichuan Province, in the mountainous areas where the production of biogas remains unsuitable, policies should concentrate on how to provide electricity to households with lower price and outage frequency, whereas in the plain and hilly areas, energy policy should focus on providing a simultaneous promotion of biogas and electricity.

References

Adamowicz, W., Louviere, J., and Williams, M. (1994). Combining revealed and stated preference methods for valuing environmental amenities. Journal of Environmental Economics and Management 26, 271–292.

Adamowicz, W., Swait, J., and Boxall, P., (1997). Perceptions versus objective measures of environmental quality in combined revealed and stated preference models of environmental valuation. Journal of Environmental Economics and Management 32, 65–84.

Ahmad, S., and Puppim de Oliveira, J.A., (2015). Fuel switching in slum and non-slum households in urban India. Journal of Cleaner Production 94, 130–136.

Ajanovic, A., (2011). Biofuels versus food production: Does biofuels production increase food prices? Energy 36, 2070–2076.

Alka, D., Neetu, M., and Vishakha, B., (2014). Biofuels: impact on food productivity, land use, environment and agriculture. International Journal of Environmental Research and Development 4(1), 9–16.

Amacher, G.S., Hyde, W.F., and Kanel, K.R., (1996). Household fuelwood demand and supply in Nepal's Tarai and Mid-Hills: choice between cash outlays and labor opportunity. World Development 24(11), 1725–1736.

An, L., Lupi, F., Liu, J., Linderman M.A., and Huang, J., (2002). Modeling the choice to switch from fuelwood to electricity: implications for giant panda habit conservation. Ecological Economics 42, 445–457.

Akpalu, W., Dasmani, I., and Aglobitse, P.B., (2011). Demand for cooking fuels in developing country: to what extent do taste and preferences matter? Energy Policy 39, 6525–6531.

Arnade, C., and Kelch, D., (2007). Estimation of Area Elasticities from a Standard Profit Function. American Journal of Agricultural Economics 89(3), 727–737.

Babcock, B.A., (2011). The impact of US biofuel policies on agricultural price levels and volatility. ICTSD Programme on Agricultural Trade and Sustainable Development Issue Paper No.35, June, 2011.

Baier, S., Clements, M., Griffiths, C., and Ihrig, J., (2009). Biofuels impact on crop and food prices: using an interactive spreadsheet. Board of Governors of the Federal Reserve System International Finance Discussion Papers Number 967, March 2009.

Baland, J.M., Bardhan, P., Das, S., Mookherjee, D., and Sarkar, R., (2010). The environmental impact of poverty: evidence from firewood collection in rural Nepal. Economic Development and Cultural Change 59(1), 23–61.

Barrett, C.B., Sherlund. S.M., and Adesina. A.A., (2008). Shadow wages, allocative inefficiency, and labor supply in smallholder agriculture. Agricultural Economics 38, 21–34.

Bebbington, A., (1999). Capitals and capabilities: a framework for analyzing peasant viability, rural livelihoods and poverty. World Development 27(12), 2021–2044.

Ben-Akiva, M., and Morikawa, T., (1990). Estimation of switching models from revealed preferences and stated intentions. Transportation Research A 24A (6), 485–495.

Benjamin, D., (1992). Household composition, labor markets, and labor demand: testing for separation in agricultural household models. Econometrica 60(2), 287–322.

Benjamin, D., and Brandt, L., (2002). Property rights, labor markets, and efficiency in a transition economy: the case of rural China. The Canadian Journal of Economics 35(4), 689–716.

Beznoska, M., (2014). Estimating a consumer demand system of energy, mobility and leisure: A microdata approach for Germany. School of Business & Economics Discussion Paper, Free University of Berlin (FU Berlin), 2014/8.

Bhattacharyya, A., and Kumbhakar, S.C., (1997). Market imperfections and output loss in the presence of expenditure constraint: A generalized shadow price approach. American Journal of Agricultural Economics 79(3), 860–871.

Biggs, E.M, Bruce, E., Boruff, B., Duncan J.M.A., Horsley, J., Pauli, N., McNeill, K., Neef, A., Van Ogtrop, F., Curnow, J., Haworth, B., Duce, S., and Imanari, Y., (2015). Sustainable development and the water-energy-food nexus: a perspective on livelihoods. Environmental Scientific Policy 54, 389–397.

Binswanger, H., (1974). A cost function approach to the measurement of factor demand and elasticities of substitution. American Journal of Agricultural Economics. 56(2), 377–386.

Blaauw, D., and Lagarde, M., (2010). Tools for implementing rural retention strategies: Towards a "how to" guide for "Discrete Choice Experiments"–A Methods Workshop", World Health Organization, Geneva, Switzerland, October 2010. Meeting report, available online at: http://www.who.int/hrh/resources/DCE_report.pdf

Bowlus, A.J., and Sicular, T., (2003). Moving towards markets? Labor allocation in rural China. Journal of Development Economics 71, 561–583.

Braun, F.G., (2010). Determinants of households' space heating type: a discrete choice analysis for German households. Energy Policy 38, 5493–5503.

Browning, M., and Chiappori, P.A., (1998). Efficient intra-houshold allocations: A general characterization and empirical test. Econometrica 66, 1241–1278.

Brownstone, D., Bunch, D.S., and Train, K., (2000). Joint mixed logit models of stated and revealed preferences for alternative-fuel vehicles. Transportation Resource Part B. 34, 315–338.

Buse, A., (1994). Evaluating the linearized almost ideal demand system. American Journal of Agricultural Economics 76, 781–793.

Byrne, J., Zhou, A., Shen, B., and Hughes, K., (2007). Evaluating the potential of small-scale renewable energy options to meet rural livelihoods needs: a GIS- and lifecycle cost-based assessment of Western China's options. Energy Policy 35(8), 4391–4401.

Cadenas, A., and Cabezudo, S., (1998). Biofuels as sustainable technologies: perspectives for less developed countries-food versus fuel? Technological Forecasting and Social Change May-June 1–2(58), 1200–1212.

Cameron, T.A., Shaw, W.D., Ragland, S.E., Callaway, J.M., and Keefe, S., (1996). Using actual and contingent behavior data with differing levels of time aggregation to model recreation demand. Journal of Agricultural Resource Economics 21, 130–149.

Carter, M.R., and Yao, Y., (2002). Local versus global separability in agricultural household models: the factor price equalization effect of land transfer rights. American Journal of Agricultural Economics 84(3), 702–715.

Castellón, C.E., Boonsaeng, T., and Carpio, C.E., (2012). Demand system estimation in the absence of price data: an application of Stone-Lewbel price indices. Selected Paper prepared for presentation at the Agricultural & Applied Economics Association's 2012 AAEA Annual Meeting, Seattle, Washington, August 12–14, 2012.

Chalfant, J., (1987) A globally flexible, almost ideal demand system. Journal of Business and Statistics 5, 233–242.

Chambers, R., and Conway, R., (1992). Sustainable rural livelihoods: practical concepts of the 21st Century. IDS Discussion Paper no. 296.

Chamdimba, O., (2009). Sustainable development of bio-energy industry in Africa. A presentation which was made to the NEPAD workshop on bio-energy that was held in CSIR. available online at: http://www.nepad.org/system/files/Renewable%20Energy%20Document-1-Oct-2009.pdf (accessed on 1 October 2009).

Charles, P., and James, M.G., (2008). Fuelwood scarcity, energy substitution and rural livelihoods in Namibia. Proceedings of the German Development Economics Conference, No.32 (2008), Zürich, Switzerland.

Chen, L., Heerink, N., and van den Berg, M., (2006). Energy consumption in rural China: a household model for three villages in Jiangxi Province. Ecological Economics 58(2), 407–420.

Chen, Y., et al., (2010). Household biogas use in rural China: A study of opportunities and constraints. Renewable and Sustainable Energy Reviews 14, 545-549.

Cherni, J.A., Dyner, I., Henao, F., Jaramillo, P., Smith, R., and Font, R.O., (2007). Energy supply for sustainable rural livelihoods. A multi-criteria decision-support system. Energy Policy 35, 1493-1504.

Cherni, J.A., and Hill, Y., (2009). Energy and policy providing for sustainable rural livelihoods in remote locations -The case of Cuba. Geoforum 40, 645-654.

Chiappori, P.A., (1988). Rational household labor supply. Econometrica 56, 63-89.

Christensen, R.G., Jorgenson, D.W., and Lau, L.J., (1973).Transcendental Logarithmic Production Frontiers. Reviews of Economics and Statistics 55, 28-45.

CNBS (China National Bureau of Statistics)., (2012). China Energy Statistical Yearbook [in Chinese].

CRES (Chinese Renewable Energy Society)., (2009). China New Energy and Renewable Energy Statistic Yearbook [in Chinese].

CRES (China Renewable Energy Society)., (2011). China New and Renewable Energy Statistic Yearbook [in Chinese].

De Janvry, A., and Sadoulet, E., (2006). Progress in the modeling of rural households' behavior under market failures. Poverty, Inequality and Development (Edited by de Janvry A. and Kanbur R.), Chapter 8, Kluwer publishing.

De Janvry, A., Fafchamps, M., and Sadoulet, E., (1991). Peasant household behavior with missing markets: Some paradoxes explained. Then Economic Journal 101(409), 1400-1417.

Deaton, A., and Muellbauer, J., (1980) Almost ideal demand system. The American Economic Review 70(3), 312-326.

Debertin, D.L., (2012). Agricultural Production Economics (Second Edition), Macmillan.

Demirbas, A.H., and Demirbas, I., (2007). Importance of rural energy bioenergy for developing countries. Energy Conversion Management 48, 2386-2398.

Démurger, S., and Fournier, M., (2011). Poverty and firewood consumption: a case study of rural households in northern China. China Economic Review 22, 512-523.

DFID., (2000). Sustainable livelihoods guidance sheets. London, UK.

Diewert, W.E., and Ostensoe, L.A., (1988). Flexible Functional Forms for Profit Functions and Global Curvature Conditions. In Barnett, W.A., Berndt, E.R., and White, H., eds., Dynamic Econometric Modeling (Cambridge: Cambridge University Press, 1988).

Diewert, W.E., and Wales, T.J., (1987). Flexible Functional Forms and Global Curvature Conditions. Econometrica 55(1), 43–68.

Diewert, W.E., and Wales, T.J., (1988). A Normalized Quadratic Semiflexible Functional Form. Journal of Econometrics 37, 327–342.

Diewert, W.E., and Wales, T.J., (1992). Quadratic Spline Models for Producer's Supply and Demand Functions. Internatinal Economic Review 33(3), 705–722.

Dutilly-Diane, C., Sadoulet, E., and de Janvry, A., (2003). Household behavior under market failures: How natural resource management in agriculture promotes livestock production in the Sahel. Journal of African Economics 12(3), 343–370.

Eales, J., and Unnevehr, L., (1988). Beef and chicken product demand. American Journal of Agricultural Economics, 70: 521–532.

Ellis, F., (2000). The determinants of rural livelihood diversification in developing countries. J. Agricultural Economics. 51(2), 289–302.

Fan, J., Liang, Y., Tao, A., Sheng, K., Ma, H., Xu, Y., Wang, C., and Sun, W., (2011). Energy policies for sustainable livelihoods and sustainable development of poor areas in China. Energy Policy 39, 1200–1212.

Farsi, M., Filippini, M., and Pachauri, S., (2007). Fuel choices in urban India households. Environmental Development Economics 12, 757–774.

Fisher, M., Shively, G.E., and Buccola, S., (2005). Activity choice, labor allocation, and forest use in Malawi. Land Economics 81(4), 503–517.

Gan, L., and Yu, J., (2008). Bioenergy transition in rural China: policy options and co-benefits. Energy Policy 36(2), 531–540.

Green, R., and Alston, J.M., (1990). Elasticities in AIDS models. American Journal of Agricultural Economics 72(2), 442–445.

Greene, W.H., and Hensher, D.A., (2010). Does scale heterogeneity across individuals matter? An empirical assessment of alternative logit models. Transportation 37, 413–428.

Green, W.H., (2012): Ecomometric Analysis. 7th ed. Upper Saddle River, NJ: Prentice Hall.

Gosens, J., Lu Y., He, G., Bluemling, B., and Beckers, T.A.M., (2013). Sustainability effects of household-scale biogas in rural China. Energy Policy 54, 273–287.

Gourieroux, C., Monfort, A., and Trognon, A., (1985). Moindres Carres Asymptotiques. Annales de l'INSEE 58, 91–122.

Gupta, C.L., (2003). Role of renewable energy technologies in generating sustainable livelihoods. Renewable and Sustainable Energy Review 7, 155–174.

Gupta, G., and Köhlin G., 2006. Preferences for domestic fuel: analysis with socio-economic factors and ranking in Kolkata, India. Ecological Economics 57(1), 107–121.

Hamermesh, D.S., (1993) Labor Demand, Princeton University Press.

Hausman, J.A, (1978). Specification tests in Econometrics. Econometrica 46, 1251–1271.

Harvey, M., and Pilgrim, S., (2011). The new competition for land: Food, energy and climate change. Food Policy 36, S40-S51

Heckman, J., (1974). Shadow Prices, Market Wages, and Labor Supply. Econometrica 42(4), 679–694.

Heckman, J., (1979). Sample selection bias as a specification error. Econometrica 47, 153–161

Heltberg, R., Arndt, T.C., and Sekhar, N.U., (2000). Fuelwood consumption and forest degradation: A household model for domestic energy substitution in rural India. Land Economics 76(2), 213–232.

Heltberg, R., (2004). Fuel switching: evidence from eight developing countries. Energy Economics 26, 869–887.

Heltberg, R., (2005). Factors determining household fuel choice in Guatemala. Environmental Development Economics 10(3), 337–361.

Henning, C. H.C.A and Henningsen, A., (2007). Modeling Farm Households' Price Responses in the Presence of Transaction Costs and Heterogeneity in Labor Market. American Journal of Agricultural Economics 89(3), 665–681.

Hensher, D.A., and Bradley, M., (1993). Using stated response choice data to enrich revealed preference discrete choice models. Market Letter 4(2), 139–151.

Hensher, D., Louviere, J., and Swait, J., (1999). Combining sources of preference data. Journal of Économics 89, 197–221.

Hensher, D.A., Greene, W.H., (2003). Mixed logit models: state of practice. Transportation 30(2), 133–176.

Hensher, D.A., Rose, J.M., and Greene, W.H., (2005). Applied Choice Analysis: A Primer. Cambridge University Press, Cambridge, MA.

Hensher, D.A., (2008). Empirical approaches to combining revealed and stated preference data: some recent developments with reference to urban mode choice. Research on Transportation Economics 23, 23–29.

Heien, D., and Wessells, C.R., (1990). Demand systems estimation with microdata: A censored regression approach. Journal of Business & Economic Statistics 8(3), 365–371.

Hosier, R.H., and Dowd, J., 1987. Household fuel choice in Zimbabwe: an empirical test of the energy ladder hypothesis. Resource and Energy 9, 347–361.

Huang, J., Haab, T.C., and Whitehead, J.C., 1997. Willingness to pay for quality improvements: should revealed and stated preference data be combined? Journal of Environmental Economics and Management 34, 240-255.

Hunsberger, C., Bolwig, S., Corbera, E., and Creutzig, F., 2014. Livelihood impacts of biofuel crop production: implications for governance. Geoforum 54, 248-260.

IEA., (2010). World Energy Outlook 2010. OECD/IEA 2010, Paris.

IIED., and ESPA., (2010). Biomass energy: Optimizing its contribution to poverty reduction and ecosystem services. Report of An International Workshop, Edinburgh, 19-21 October 2010.

Jacoby, H., (1992). Productivity of men and women and the sexual division of labor in peasant agriculture of the Peruvian Sierra. Journal of Development Economics 37, 265-287.

Jacoby, H., (1993). Shadow wages and peasant family labor supply: An econometric application to the Peruvian Sierra. The Review of Economic Studies 60(4), 903-921

Jiang, L., and O'Neill, B.C., (2004). The energy transition in rural China. International Journal of Global Energy Issues 21(1/2), 2-26.

Joshee, B.R., Amacher, G.S., and Hyde, W.F., (2000). Household fuel production and consumption, substitution, and innovation in two districts of Nepal. Economics of Forestry and Rural Development. An Empirical Introduction from Asia (Edited by Hyde W.F., Amacher G.S., and Colleagues), Chapter 4, The University of Michigan Press.

Just, R.E., Zilberman, D., and Hochman, E., (1983). Estimation of multicrop production functions. American Journal of Agricultural Economics 65(4), 770-780.

Kanagawa, M., Nakata, T., (2007). Analysis of the energy access improvement and its socio-economic impacts in rural areas of developing countries. Ecological Economics 62, 319-329.

Kaygusuz, K., (2011). Energy services and energy poverty for sustainable rural development. Renewable and Sustainable Energy Reviews 15, 936-947.

Key, N., Sadoulet, E., and de Janvry, A., (2000). Transaction costs and agricultural household supply response. American Journal of Agricultural Economics 82, 245-259.

Kgathi, D.C., (1997). Biomass Energy Policy in Africa: Selected Case Studies. United Kingdom: Pergaman Press.

Kgathi, D.L., and Mfundisi, K.B., (2009). Potential impacts of the production of liquid biofuels on food security in Botswana. Report Submitted to the

COMPETE Project as a Contribution to the Task of Work Package One (WP1) Activities: Current Land Use Patterns and Impacts, September, 2009.

Kodde, D.A., Palm, F.C., and Pfann, G.A., (1990), Asymptotic Least-Squares Estimation Efficiency Considerations and Applications. Journal of Applied Econometrics 5, 229–243.

Koebel, B., Falk, M. and Laisney, F., (2000). Imposing and Testing Curvature Conditions on a Box-Cox Cost Function. Discussion Paper No. 00–70, ZEW, Mannheim, Germany.

Koebel, B., Falk, M. and Laisney, F., (2003). Imposing and Testing Curvature Conditions on a Box-Cox Cost Function. Journal of Business and Economic Statistics 21 (2), 319–335.

Köhlin, G., and Parks, P.J., (2001). Spatial variability and disincentives to harvest: Deforestation and fuelwood collection in South Asia. Land Economics 77(2), 206–218.

Kohli, U., (1993). A Symmetric Normalized Quadratic Profit GNP Function and the U.S. Demand for Imports and Supply of Exports. International Economic Review 34(1), 243–255.

Kowsari, R., Zerriffi, H., (2011). Three dimensional energy profile: a conceptual framework for assessing household energy use. Energy Policy 39, 7505–7517.

Kroes, E.P., and Sheldon, R.J., (1988). Stated preference methods: an introduction. Journal of Transportation Economic Policy 22(1), 11–25.

Kusago, T., and Barham, B. L., (2001). Preference heterogeneity, power, and intrahousehold decision-making in rural Malaysia. World Development 29(7), 1237–1256.

Kwakwa, P.A., Wiafe, E.D., and Alhassan, H., (2013). Households energy choice in Ghana. Journal of Empirical Economics 1(3), 96–103.

Lagarde, M., and Blaauw, D., (2009). A review of the application and contribution of discrete choice experiments to inform human resources policy interventions. Human Resources for Health, 7: 62. available online at: http://www.human-resources-health.com/content/7/1/62 (accessed on 24 July 2009)

Lamb, R.L., (2001). Fertilizer use, risk and off-farm labor markets in the Semi-Arid Tropics of India. Department of Agricultural and Resource Economics Report No.23. North Carolina State University, U.S.A.

Lau, L.J., (1972). Profit Functions of Technologies with Multiple Inputs and Outputs. The Review of Economics and Statistics 54(3), 281–289.

Lau, L.J., (1978). Testing and Imposing Monotonicity, Convexity and Quasiconvexity Constraint. In Fuss, M., and McFadden, D., eds., Production Economics: A Dual Approach to Theory and Applications, Vol.1 (Amsterdam: North-Holland, 1978).

Leach, G., (1987). Household Energy Handbook: an Interim Guide and Reference Manual. World Bank, Washington D.C.

Leach, G., (1992). The energy transition. Energy Policy 20(2), 116–123.

Le, K.T., (2010). Separation hypothesis tests in the agricultural household model. American Journal of Agricultural Economics 92(5), 1420–1431.

Le, K.T., (2009). Shadow wages and shadow income in farmers' labor supply functions. American Journal of Agricultural Economics 91(3), 685–696.

Lee, S.M., Kim, Y.S., Jaung, W., Latifah, S., Afifi, M., and Fisher, L.A., (2015). Forests, fuelwood and livelihoods - energy transition patterns in eastern Indonesia. Energy Policy 85, 61–70.

Lewbel, A., (1989). Identification and estimation of equivalence scales under weak separability. The Review of Economic Studies 62, 311–316.

Li, J.J., Zhuang, X., DeLaquil, P., and Larson, E.D (2011). Biomass Energy in China and Its Potential. Energy for Sustainable Development 4(4), 66–80.

Lopez, R.E., (1984). Estimating labor supply and production decisions of self-employed farm producers. European Economic Review 24, 61–82.

Louviere, J.J., Hensher, D.A., Swait, J.D., (2000). Stated Choice Methods: Analysis and Application. Cambridge University Press. Cambridge, MA.

Louviere, J.J., Hensher, D.A., (1983). Using discrete choice models with experimental data to forecast consumer demand for a unique cultural event. Journal of Consumer Research 10, 348–361.

MaCurdy, T.E., and Pencavel, J.H., (1986). Testing between competing models of wage and employment determination in unionized markets. Journal of Political Economy 94(3), S3–39.

Manser, M., and Brown, M., (1980). Marriage and household decision-making: a bargaining analysis. International Economic Review 21, 31–44.

Manski, C.F., (2001). Daniel McFadden and the econometric analysis of discrete choice. The Scand. Journal of Economics 103(2), 217–230.

Mark, T.L., Swait, J., (2004). Using stated preference and revealed preference modeling to evaluate prescribing decisions. Health Economics 13: 563–573.

Mas-colell, A., Whinston, M.D., and Green, J., (1995). Microeconomic Theory. Oxford Student Edition, Oxford University Press.

Masera, O.R., Saatkamp, B.D., Kammen, D.M., (2000). From linear fuel switching to multiple cooking strategies: a critique and alternative to the energy ladder model. World Development. 28(12), 2083–2103.

Mathenge, M.K., and Tschirley, D. (2007). Off-farm and farm production decisions: evidence from maize-producing households in rural Kenya. Paper submitted for the CSAE Conference 2007 on 'Economic Development in Africa', St. Catherine's College, University of Oxford, UK. March 18-20, 2007.

Mathye, R., (2002). Environmental and Socio-economic Impacts of Biomass Energy Consumption in the Mbhokota Village, Northern Province. Mini-thesis submitted in partial fulfilment of the requirements for the degree Master of Arts in Geography and Environmental Management at the Rand Afrikaans University.

McFadden, D., (1973). Conditional logit analysis of qualitative choice behavior, in: Zaremmbka, P. (Ed.), Frontiers in Econometrics. Academic Press, New York, pp. 105-142.

McFadden, D., (1974). The measurement of urban travel demand. Journal of Public Economics 3(4), 303-328

McFadden, D., (1980). Econometric models for probabilistic choice among products. The Journal of Business 53(3) Part.2: Interfaces between Marketing and Economics, S13-S29.

Meisen, P., and Cavino, N., (2007). Rural Electricification, Human Development and the Renewable Energy Potential of China. Global Energy network Institute, October, 2007.

Mekonnen, A., (1999). Rural household biomass fuel production and consumption in Ethiopia: a case study. Journal of Forest Economics 5(1), 69-97.

Mekonnen, A., and Köhlin, G., (2008). Determinants of household fuel choice in major cities in Ethiopia. Environment for Development Discussion Paper Series EfD DP 08-18.

Mishra, A., (2008). Determinants of fuelwood use in rural Orissa: Implications for energy transition. Working Paper of SANDEE, No.37-08.

Mittelhammer, R.C., Matulich, S.C., and Bushaw, D., (1981). On implicit forms of multiproduct-multifactor production functions. American Journal of Agricultural Economics 63(1), 164-168.

MOA (Ministry of Agricultural of China), (2007a). National Rural Biogas Construction Plan (2006-2010). [in Chinese]

MOA (Ministry of Agricultural of China), (2007b).Development Plan for the Agricultural Bioenergy Industry (2007-2015). [in Chinese]

Morikawa, T., (1989). Incorporating stated preference data in travel demand analysis. Submitted to the Department of Civil Engineering in Partial Fulfillment of the Requirement for the Degree of Doctor of Philosophy at the Massachusetts Institute of Technology. June 1989.

Morikawa, T., Ben-Akiva, M., and Yamada, K., (1991). Forecasting intercity rail ridership using revealed preference and stated preference data. Transportation Research Record 1328, 30-35.

Muwaura, F et al., (2014). Determinants of household's choice of cooking energy in Uganda. EPRC Research Series no.114.

NDRC (National Development and Reform Commission)., (1997). China's Agenda 21-White paper on China's population, environment, and development in 21st Century, Science and Technology Outline for Sustainable Development. The People's Republic of China National Report on Sustainable Development. [in Chinese]

NDRC (National Development and Reform Commission)., (2002). China's Agenda 21-White paper on China's population, environment, and development in 21st Century, Science and Technology Outline for Sustainable Development. The People's Republic of China National Report on Sustainable Development. [in Chinese]

NDRC (National Development and Reform Commission)., (2004). China National Energy Strategy and Policy 2020. [in Chinese]

NDRC (National Development and Reform Commission)., (2007). Medium and long-term development plan for renewable energy in China. [in Chinese]

NRDC/MOA., (2011). Announcement on the Strengthening of Rural Biogas Construction. First Tranche of 2011. [in Chinese]

Nepal, K.P., Fukuda, D., and Yai, T., (2005). Microeconomic models of intra-household activity time allocations. Journal of Eastern Society for Transportation Studies 6, 1637–1650.

Nyang, F.O., (1999). Household energy demand and environmental management in Kenya: An Application of the Agricultural Household Model. PhD dissertation, Chapter 8, University of Amsterdam.

Ouedraogo, B., (2006). Household energy preferences for cooking in urban Ouagadougou, Burkina Faso. Energy Policy 34, 3787–3795.

Peng, W., Hisham, Z., and Pan, J., (2010). Household level fuel switching in rural Hubei. Energy for Sustainable Development 14, 238–244.

Ping, X.G. et al., (2012). Social and ecological effects of biomass utilization and the willingness to use clean energy in the eastern Qinghai-Tibet Plateau. Energy Policy 51, 828–833.

Qu, W., Tu, Q., and Bluemling, B., (2013). Which factors are effective for farmers' biogas use?-Evidence from a large-scale survey in China. Energy Policy 63, 26–33.

Ramji, A et al., (2012). Rural energy access and inequalities: An analysis of NSS data from 1999–00 to 2009–10. TERI-NFA Working Paper No. 4, December, 2012.

Rao, K.D., Shroff, Z., Ramani, S., Khandpur, N., Murthy, S., Hazarika, I., Choksi, M., Ryan, M., Berman, P., and Vujicic, M., (2012). How to attract health workers to rural areas? Findings from a discrete choice experiment from India. HNP Discussion Paper Series no 74544. World Bank, Washington, D.C.

Reddy, B.S., (1995). A multilogit model for fuel shifts in the domestic sector. Energy 20(9), 929–936.

Sadoulet, E., and de Janvry, A., (1995). Quantitative Development Policy Analysis. The Johns Hopkins University Press.

Schlag, N., and Zuzarte, F., (2008). Market barriers to clean cooking fuels in Sub-Saharan Africa: A review of literature. An SEI Working Paper from Stockholm Environment Institute, Sweden.

Schneider, D., (2011). The labor share: a review of theory and evidence. SFB 649 Discussion Paper 2011–069, Humdoldt-Universitaet zu Berlin.

SCREO (Sichuan Rural Energy Office), (2013). Compilation of Sichuan Rural Renewable Energy Statistics (2013). [in Chinese]

Scoones, I., (2009). Livelihoods perspectives and rural development. Journal of Peasant Study 36(1), 171–196.

Shen, J., Saijo, T., (2009). Does an energy efficiency label alter consumers' purchasing decisions? a latent class approach based on a stated choice experiment in Shanghai. Journal of Environmental Management 90, 3561–3573.

Shively, G.E., (2001). Agricultural change, rural labor markets, and forest clearing: An illustrative case from Philippines. Land Economics 77(2), 268–284.

Shumway, C.R., Pope, R.D., and Nash, E.K., (1984). Allocable fixed inputs and jointness in agricultural production: Implications for economic modeling. American Journal of Agricultural Economics, 72–78.

Singh, I., Squire, L., and Strauss, J., (1986). Agricultural household models: Extensions, applications and policy. A World Bank Research Publication, The Johns Hopkins University Press.

Skoufias, E., (1993). Labor market opportunities and intrafamily time allocation in rural households in South Asia. Journal of Development Economics 40, 277–310.

Skoufias, E., (1994). Using shadow wages to estimate labor supply of agricultural households. American Journal of Agricultural Economics 76, 215–227.

Sokhansanj, S., Cushman, J., and Wright, L., (2003). Collection and Delivery of Biomass for Fuel and Power Production. Agricultural Engineering International: the CIGR Journal of Scientific Research and Development. Invited Overview Paper. Vol. V. February 2003.

Soloaga, I., (1999). The treatment of non-essential inputs in Cobb-Douglas technology: An application to Mexican rural household-level data. Policy Research Working Paper, World Bank, December 1999.

Spautz, L., Charron, D., Dunaway, J., Hao, F., and Chen, X., (2006). Spreading innovative biomass stove technologies through China and beyond. Boiling Point 52, 6–8.

Spencer, D.E., and Berk, K.N., (1981). A limited information specification test. Econometrica 49(4), 1079–1085.

Suharo, A., (2002). An Almost Ideal Demand System for Food Based on Cross Section Data: Rural and Urban East Java, Indonesia. Doctoral Dissertation, Faculty of Agricultural science, University of Goettingen.

Suliman, K.M., (2013). Factors affecting the choice of households' primary cooking fuel in Sudan. Working Paper 760, Department of Economics, University of Khartoum, Sudan.

Swait, J., Louviere, J.J., and Williams, M., (1994). A sequential approach to exploiting the combined strengths of SP and RP data: application to freight shipper choice. Transportation 21(2), 135–152.

Takama, T., Lambe, F., Johnson, F.X., Arvidson, A., Atanassov, B., Debebe, M., Nilsson, L., Tella, P., and Tsephel, S., (2011). Will African consumers buy cleaner fuels and stoves? a household energy economic analysis model for the market introduction of bio-ethanol cooking stoves in Ethiopia, Tanzania, and Mozambique. SEI Research Report. Stockholm Institute, Stockholm, Sweden. ISBN 978-91-86125-25-7.

Takama, T., Tsephel, S., and Johnson, F.X.., (2012). Evaluating the relative strength of product-specific factors in fuel switching and stove choice decisions in Ethiopia. A discrete choice model of household preferences for clean cooking alternatives. Energy Econ. 34, 1763–1773.

Tiberti, L., and Tiberti, M., (2015). Rural Policies, Price Change and Poverty in Tanzania: an Agricultural Household Model-based Assessment. CIRPÉE Working Paper 12-29.

Timilsina, G.R., Begbin, J.C., van der Mensbrugghe, D., and Mevel, S., (2010). The impacts of biofuel targets on land-use change and food supply: A global CGE assessment. Policy Research Working Paper 5513. The World Bank Development Research Group Environment and Energy Team, December 2010.

Train, K., (2003). Discrete choice models with simulation. Cambridge University Press.

Vaage, K., (2000). Heating technology and energy use: a discrete/continuous choice approach to Norwegian household energy demand. Energy Econ. 22, 649–666.

Vakis, R., Sadoulet, E., de Janvry, A., and Cafiero, C., (2004). Testing for separability in household models with heterogeneous behavior: A mixture model approach. CUDARE Working Papers. Available at: http://escholarship.org/uc/item/4hs3g5dj.

Van der Kroon, B., Brouwer, R., and van Beukering, P.J.H., (2013). The energy ladder: theoretical myth or empirical truth? results from a Meta-analysis. Renewable and Sustainable Energy Review 20, 504–513.

Van Horen, C., and Eberhard, A.A., (1995). Energy, Environment and the Rural Poor in South Africa. Development Southern Africa. 12(2), 197–211.

Varian., (1978). Micoeconomic Analysis. New York: W.W.Norton & Company.

Villezca-Becerra, P.A., and Schumway, C.R., (1992). Multiple-output Production Modeled with Three Functional Forms. Journal of Agricultural and Resource Economics 17(1), 13–28.

Von Lampe, M., (2007). Economics and agricultural market impacts of growing biofuel production. Agrarwirtschaft 56, Heft 5/6, 232–237.

Wambua, S.M., (2011). Household energy consumption and dependency on common pool forest resources: the case of Kakamega forest, Western Kenya. Dissertation to obtain the Ph.D.degree in the International Ph.D. Program for Agricultural Sciences in Göttingen at the Faculty of Agricultural Sciences, Georg-August-University Göttingen, Germany.

Wang, C.C. et al,. (2012). Rural household livelihood change, fuelwood substitution, and hilly ecosystem restoration: Evidence from China. Renewable and Sustainable Energy Reviews 16, 2475–2482.

Wang, Q., and Qiu, H., (2009). Prevention of Tibetan eco-environmental degradation caused by traditional use of biomass. Energy Reviews 13, 2181–2186.

Wang, X., Herzfeld, T., and Glauben, T., (2007). Labor allocation in transition: evidence from Chinese rural households. China Economic Review 18, 287–308.

Weaver, R.D., (1983). Multiple input, multiple output production choices and technology in the U.S. wheat region. American Journal of Agricultural Economics 65(1), 45–56.

West, S.E., and Parry, I.W.H., (2009). Alcohol/leisure complementarities: Empirical estimates and implications for tax policy. Working Paper of Resources for the Future, (March, 2009) RFF DP 09–09.

WHO, (2006). Fuel for life: household energy and health. Printed in France.

WHO, (2012). How to conduct a discrete choice experiment for health workforce recruitment and retention in remote and rural areas: a user guide with case studies. Printed in France.

Whitehead J.C., Pattanayak S.K., Van Houtven G.L., and Gelso B.R., (2008). Combining revealed and stated preference data to estimate the nonmarket value of ecological service: an assessment of the state of the science. Journal of Economic Survey 22(5), 872–908.

Wiedenmann, R., (1991). Deforestation from the overexploitation of wood resources as a cooking fuel: A comment on the optimal control model of Hassan and Hertzler. Energy Economics 13(2), 81–85.

Willis, K., Scarpa, R., Gilroy, R., and Hamza, N., (2011). Renewable energy adoption in an aging population: heterogeneity in preferences for micro-generation technology adoption. Energy Policy 39, 6021–6029.

World Bank (WDB)., World Bank Open Data Initiative Datasets of Rural Population in China. Available at: http://data.worldbank.org/indicator/SP.RUR.TOTL?page=1

Yong, M., (2015). What's the future for Sichuan household biogas digester construction in transition? Daily Newspaper of Rural Sichuan, 17th, July, 2015. [in Chinese] Available at: http://r.m.baidu.com/5k7pvk3

Yu, B., Zhang, J., and Fujiwara, A., (2012). Analysis of the residential location choice and household energy consumption behavior by incorporating multiple self-selection effects. Energy Policy 46, 319–334.

Zafar, S., (2015). Collection Systems for Agricultural Biomass. Available online at (access at June 25, 2015): http://www.bioenergyconsult.com/biomass-collection/.

Zellner, A., (1962). An efficient method of estimating seemingly unrelated regressions and tests for aggregation bias. Journal of the American Statistical Association 57, 348–368.

Zellner, A., (1963). Estimators for seemingly unrelated regression equations: Some exact finite sample sesults. Journal of the American Statistical Association 58, 977–992.

Zellner, A., and Huang D.S., (1962). Further properties of efficient estimators for seemingly unrelated regression equations. International Economic Review 3, 300–313.

Zhang, L.X., et al., (2009). Rural energy in China: Pattern and Policy. Renewable Energy 34, 2813–2823.

Zhang, Q., Watanabe, M., Lin, Tun., Delaquil, P., Wang, G., and Alipalo, M.H., (2010). Rural biomass energy 2020: Cleaner Energy, Better Environment, Higher Rural Income, People's Republic of China. Asia Development Bank, Mandaluyong City, Philippines.

Zilberman, D., Hochman, G., Rajagopal, D., Sexton, S., and Timilsina, G., (2013). The impact of biofuels on commodity food prices: Assessment of findings. American Journal of Agricultural Economics 95(2), 275–281.

Appendix

Figure A.1 Optimal labor allocation in separable AHM

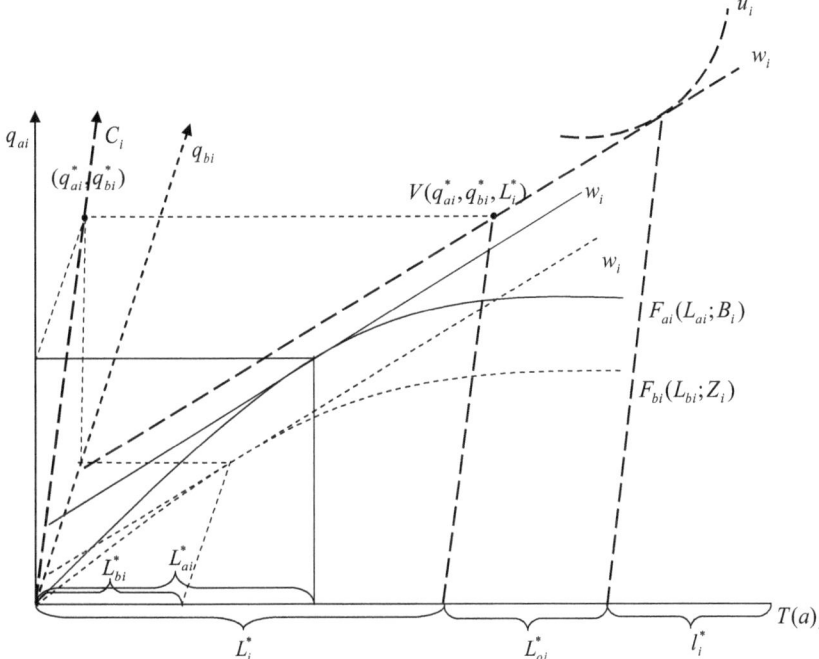

Source: Author's depiction

Figure A.2 The changes in equilibrium with decrease in wage rate in separable AHM

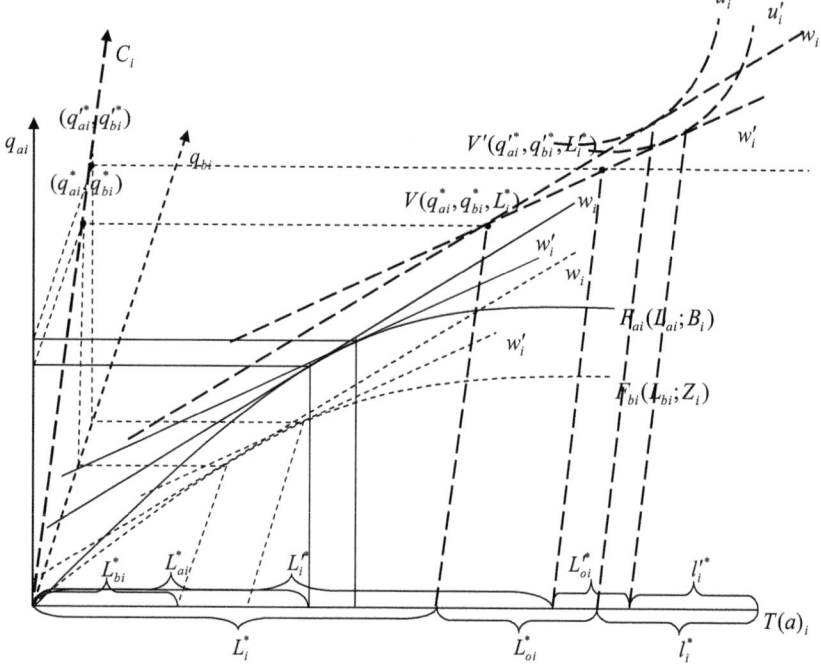

Source: Author's depiction

Figure A.3 The changes in equilibrium with increase in wage rate in separable AHM

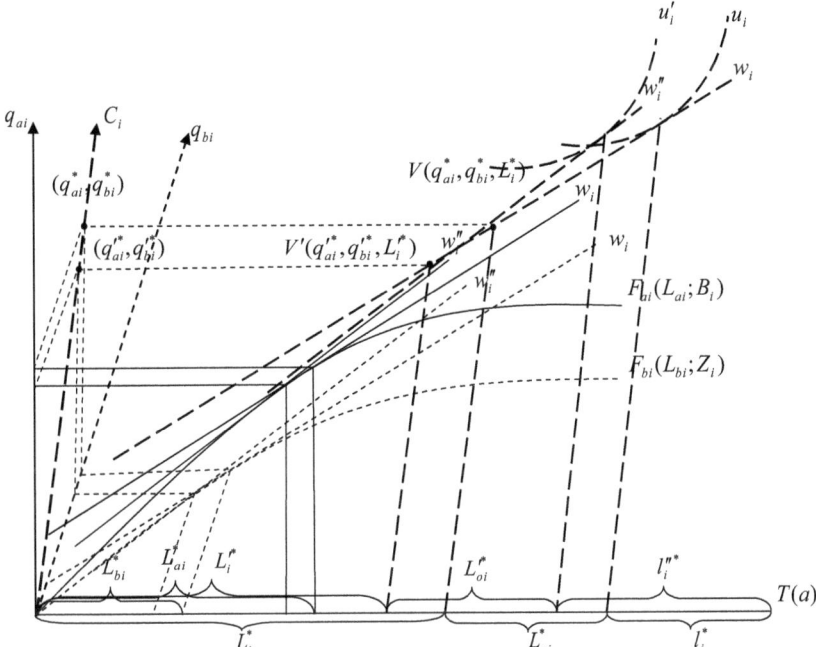

Source: Author's depiction

Figure A.4 Optimal labor allocation in non-separable AHM

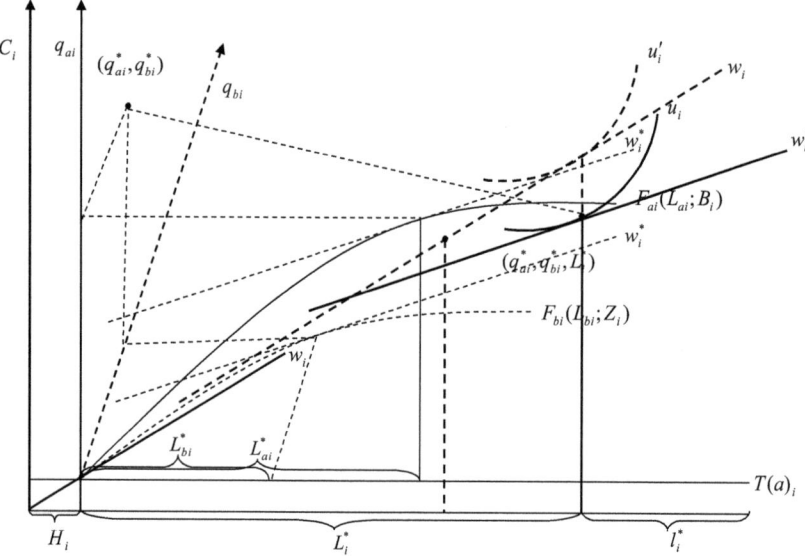

Source: Author's depiction

Table A.1 Estimation results of the normalized quadratic profit function with imposition of convexity

Parameter	Coef.^{u}	T-Stat	Coef.^{r}
α_a	37948.0339**	2.2065	50139.6383
α_b	−43.4072	(−0.0365)	1421.9070
α_l	−2104.5469	(−1.1996)	−4206.8905
α_o	−5585.7248***	(−4.1370)	−7045.2400
$\beta_{aa}(p_a p_a)$	−2202.7377**	(−2.3939)	2917.0882
$\beta_{ab}(p_a p_b^*)$	−1536.2492***	(−2.7964)	−1457.4017
$\beta_{al}(p_a p_l^*)$	2605.3060***	5.2161	−203.6292
$\beta_{ao}(p_a p_o)$	1133.6809**	2.4797	−1256.0572
$\beta_{bb}(p_b^* p_b^*)$	−429.1761	(−2.7374)	795.8768
$\beta_{bl}(p_b^* p_l^*)$	1146.4071***	6.2630	275.2318
$\beta_{bo}(p_b^* p_o)$	819.0182***	4.0979	386.2931
$\beta_{ll}(p_l^* p_l^*)$	−1648.3134***	(−5.0531)	458.5478
$\beta_{lo}(p_l^* p_o)$	−2103.3997***	(−7.8839)	−530.1505
$\beta_{oo}(p_o p_o)$	150.7006	0.5777	1399.9146
$\delta_{aAL}(z_{AL})$	6679.3614	1.3957	8125.4508
$\delta_{bAL}(z_{AL})$	42.7095	0.1323	182.6015
$\delta_{lAL}(z_{AL})$	−806.2899*	−1.6694	−1099.5683
$\delta_{oAL}(z_{AL})$	408.7647	1.1021	183.8917
$\gamma_{aALAL}(p_a z_{AL} z_{AL})$	−402.0526	(−1.0679)	−467.6394
$\gamma_{bALAL}(p_b^* z_{AL} z_{AL})$	6.5346	0.2597	0.1591
$\gamma_{lALAL}(p_l^* z_{AL} z_{AL})$	41.7793	1.1032	54.2255
$\gamma_{oALAL}(p_o z_{AL} z_{AL})$	−14.9734	(−0.5167)	−5.4900
Hausman test statistic		38.94***	
No. of Obs.		556	

Note: The system of SNQ profit function and netput equations are estimated using R package 'micEconSNQP'. The significance levels are: *10%, **5%, ***1%. The missing dummy for regions is Hilly area. The superscript 'u' refers to the estimated coefficients of unrestricted profit function, whereas 'r' is the ones of restricted estimation. T-Stat refers to the estimate parameter to the left. Subscript 'a' represents agricultural outputs, 'b' denotes amount of collected biomass, 'l' is labor inputs and 'o' refers to intermediate inputs.

Development Economics and Policy

Series edited by Franz Heidhues†, Joachim von Braun, Ulrike Grote and Manfred Zeller

Vol. 1 Andrea Fadani: Agricultural Price Policy and Export and Food Production in Cameroon. A Farming Systems Analysis of Pricing Policies. The Case of Coffee-Based Farming Systems. 1999.

Vol. 2 Heike Michelsen: Auswirkungen der Währungsunion auf den Strukturanpassungsprozeß der Länder der afrikanischen Franc-Zone. 1995.

Vol. 3 Stephan Bea: Direktinvestitionen in Entwicklungsländern. Auswirkungen von Stabilisierungsmaßnahmen und Strukturreformen in Mexiko. 1995.

Vol. 4 Franz Heidhues / François Kamajou: Agricultural Policy Analysis – Proceedings of an International Seminar, held at the University of Dschang, Cameroon on May 26 and 27 1994, funded by the European Union under the Science and Technology Program (STD). 1996.

Vol. 5 Elke M. Förster: Protection or Liberalization? A Policy Analysis of the Korean Beef Sector. 1996.

Vol. 6 Gertrud Schrieder: The Role of Rural Finance for Food Security of the Poor in Cameroon. 1996.

Vol. 7 Nestor R. Ahoyo Adjovi: Economie des Systèmes de Production intégrant la Culture de Riz au Sud du Bénin: Potentialités, Contraintes et Perspectives. 1996.

Vol. 8 Jenny Müller: Income Distribution in the Agricultural Sector of Thailand. Empirical Analysis and Policy Options. 1996.

Vol. 9 Michael Brüntrup: Agricultural Price Policy and its Impact on Production, Income, Employment and the Adoption of Innovations. A Farming Systems Based Analysis of Cotton Policy in Northern Benin. 1997.

Vol. 10 Justin Bomda: Déterminants de l'Epargne et du Crédit, et leurs Implications pour le Développement du Système Financier Rural au Cameroun. 1998.

Vol. 11 John M. Msuya: Nutrition Improvement Projects in Tanzania: Implementation, Determinants of Performance, and Policy Implications. 1998.

Vol. 12 Andreas Neef: Auswirkungen von Bodenrechtswandel auf Ressourcennutzung und wirtschaftliches Verhalten von Kleinbauern in Niger und Benin. 1999.

Vol. 13 Susanna Wolf (ed.): The Future of EU-ACP Relations. 1999.

Vol. 14 Franz Heidhues / Gertrud Schrieder (eds.): Romania – Rural Finance in Transition Economies. 2000.

Vol. 15 Katinka Weinberger: Women's Participation. An Economic Analysis in Rural Chad and Pakistan. 2000.

Vol. 16 Christof Batzlen: Migration and Economic Development. Remittances and Investments in South Asia: A Case Study of Pakistan. 2000.

Vol. 17 Matin Qaim: Potential Impacts of Crop Biotechnology in Developing Countries. 2000.

Vol. 18 Jean Senahoun: Programmes d'ajustement structurel, sécurité alimentaire et durabilité agricole. Une approche d'analyse intégrée, appliquée au Bénin. 2001.

Vol. 19 Torsten Feldbrügge: Economics of Emergency Relief Management in Developing Countries. With Case Studies on Food Relief in Angola and Mozambique. 2001.

Vol. 20 Claudia Ringler: Optimal Allocation and Use of Water Resources in the Mekong River Basin: Multi-Country and Intersectoral Analyses. 2001.

Vol. 21 Arnim Kuhn: Handelskosten und regionale (Des-)Integration. Russlands Agrarmärkte in der Transformation. 2001.

Vol. 22 Ortrun Anne Gronski: Stock Markets and Economic Growth. Evidence from South Africa. 2001.

Vol. 23 Patrick Webb / Katinka Weinberger (eds.): Women Farmers. Enhancing Rights, Recognition and Productivity. 2001.

Vol. 24 Mingzhi Sheng: Lebensmittelkonsum und -konsumtrends in China. Eine empirische Analyse auf der Basis ökonometrischer Nachfragemodelle. 2002.

Vol. 25 Maria Iskandarani: Economics of Household Water Security in Jordan. 2002.

Vol. 26 Romeo Bertolini: Telecommunication Services in Sub-Saharan Africa. An Analysis of Access and Use in the Southern Volta Region in Ghana. 2002.

Vol. 27 Dietrich Müller-Falcke: Use and Impact of Information and Communication Technologies in Developing Countries' Small Businesses. Evidence from Indian Small Scale Industry. 2002.

Vol. 28 Wolfram Erhardt: Financial Markets for Small Enterprises in Urban and Rural Northern Thailand. Empirical Analysis on the Demand for and Supply of Financial Services, with Particular Emphasis on the Determinants of Credit Access and Borrower Transaction Costs. 2002.

Vol. 29 Wensheng Wang: The Impact of Information and Communication Technologies on Farm Households in China. 2002.

Vol. 30 Shyamal K. Chowdhury: Institutional and Welfare Aspects of the Provision and Use of Information and Communication Technologies in the Rural Areas of Bangladesh and Peru. 2002.

Vol. 31 Annette Luibrand: Transition in Vietnam. Impact of the Rural Reform Process on an Ethnic Minority. 2002.

Vol. 32 Felix Ankomah Asante: Economic Analysis of Decentralisation in Rural Ghana. 2003.

Vol. 33 Chodechai Suwanaporn: Determinants of Bank Lending in Thailand: An Empirical Examination for the Years 1992 to 1996. 2003.

Vol. 34 Abay Asfaw: Costs of Illness, Demand for Medical Care, and the Prospect of Community Health Insurance Schemes in the Rural Areas of Ethiopia. 2003.

Vol. 35 Gi-Soon Song: The Impact of Information and Communication Technologies (ICTs) on Rural Households. A Holistic Approach Applied to the Case of Lao People's Democratic Re- public. 2003.

Vol. 36 Daniela Lohlein: An Economic Analysis of Public Good Provision in Rural Russia. The Case of Education and Health Care. 2003.

Vol. 37 Johannes Woelcke. Bio-Economics of Sustainable Land Management in Uganda. 2003.

Vol. 38 Susanne M. Ziemek: The Economics of Volunteer Labor Supply. An Application to Countries of a Different Development Level. 2003.

Vol. 39 Doris Wiesmann: An International Nutrition Index. Concept and Analyses of Food Insecurity and Undernutrition at Country Levels. 2004.

Vol. 40 Isaac Osei-Akoto: The Economics of Rural Health Insurance. The Effects of Formal and Informal Risk-Sharing Schemes in Ghana. 2004.

Vol. 41 Yuansheng Jiang: Health Insurance Demand and Health Risk Management in Rural China. 2004.

Vol. 42 Roukayatou Zimmermann: Biotechnology and Value-added Traits in Food Crops: Relevance for Developing Countries and Economic Analyses. 2004.

Vol. 43 F. Markus Kaiser: Incentives in Community-based Health Insurance Schemes. 2004.

Vol. 44 Thomas Herzfeld: *Corruption begets Corruption*. Zur Dynamik und Persistenz der Korruption. 2004.

Vol. 45 Edilegnaw Wale Zegeye: The Economics of On-Farm Conservation of Crop Diversity in Ethiopia: Incentives, Attribute Preferences and Opportunity Costs of Maintaining Local Varieties of Crops. 2004.

Vol. 46 Adama Konseiga: Regional Integration Beyond the Traditional Trade Benefits: Labor Mobility contribution. The Case of Burkina Faso and Côte d'Ivoire. 2005.

Vol. 47 Beyene Tadesse Ferenji: The Impact of Policy Reform and Institutional Transformation on Agricultural Performance. An Economic Study of Ethiopian Agriculture. 2005.

Vol. 48 Sabine Daude: Agricultural Trade Liberalization in the WTO and Its Poverty Implications. A Study of Rural Households in Northern Vietnam. 2005.

Vol. 49 Kadir Osman Gyasi: Determinants of Success of Collective Action on Local Commons. An Empirical Analysis of Community-Based Irrigation Management in Northern Ghana. 2005.

Vol. 50 Borbala E. Balint: Determinants of Commercial Orientation and Sustainability of Agricultural Production of the Individual Farms in Romania. 2006.

Vol. 51 Pamela Marinda: Effects of Gender Inequality in Resource Ownership and Access on Household Welfare and Food Security in Kenya. A Case Study of West Pokot District. 2006.

Vol. 52 Charles Palmer: The Outcomes and their Determinants from Community-Company Contracting over Forest Use in Post-Decentralization Indonesia. 2006.

Vol. 53 Hardwick Tchale: Agricultural Policy and Soil Fertility Management in the Maize-based Smallholder Farming System in Malawi. 2006.

Vol. 54 John Kedi Mduma: Rural Off-Farm Employment and its Effects on Adoption of Labor Intensive Soil Conserving Measures in Tanzania. 2006.

Vol. 55 Mareike Meyn: The Impact of EU Free Trade Agreements on Economic Development and Regional Integration in Southern Africa. The Example of EU-SACU Trade Relations. 2006.

Vol. 56 Clemens Breisinger: Modelling Infrastructure Investments, Growth and Poverty Impact. A Two-Region Computable General Equilibrium Perspective on Vietnam. 2006.

Vol. 57 Meike Wollni: Coping with the Coffee Crisis. An Analysis of the Production and Marketing Performance of Coffee Farmers in Costa Rica. 2007.

Vol. 58 Franklin Simtowe: Performance and Impact of Microfinance. Evidence from Joint Liability Lending Programs in Malawi. 2008.

Vol. 59 Xiangping Jia: Credit Rationing and Institutional Constraint. Evidence from Rural China. 2008.

Vol. 60 Holger Seebens: The Economics of Gender and the Household in Developing Countries. 2009.

Vol. 61 Stephan Piotrowski: Land Property Rights and Natural Resource Use. An Analysis of Household Behavior in Rural China. 2009.

Vol. 62 Sebastian M. Scholz: Rural Development through Carbon Finance. Forestry Projects under the Clean Development Mechanism of the Kyoto Protocol. Assessing Smallholder Participation by Structural Equation Modeling. 2009.

Vol. 63 Jakob Rupert Friederichsen: Opening Up Knowledge Production through Participatory Research? Agricultural Research for Vietnam's Northern Uplands. 2009.

Vol. 64 Olivier Ecker: Economics of Micronutrient Malnutrition. The Demand for Nutrients in Sub-Saharan Africa. 2009.

Vol. 65 Julia Johannsen: Operational Assessment of Monetary Poverty by Proxy Means Tests. 2009

Vol. 66 Ephraim Nkonya / Nicolas Gerber / Philipp Baumgartner / Joachim von Braun / Alessandro De Pinto / Valerie Graw / Edward Kato / Julia Kloos / Teresa Walter: The Economics of Land Degradation. Toward an Integrated Global Assessment. 2011.

Vol. 67 S. Idriss Nazaire Houssou: Operational Poverty Targeting by Proxy Means Tests. Models and Policy Simulations for Malawi. 2013.

Vol. 68 Abdul Salam Lodhi: Education, Child Labor and Human Capital Formation in Selected Urban and Rural Settings of Pakistan. 2013.

Vol. 69 Evita Hanie Pangaribowo: Household Food Consumption, Women´s Asset and Food Policy in Indonesia. 2013.

Vol. 70 Dan Liu: China's New Rural Cooperative Medical Scheme. Evolution, Design and Impacts. 2013.

Vol. 71 Camille Saint-Macary: Microeconomic Impacts of Institutional Change in Vietnam's Northern Uplands. Empirical Studies on Social Capital, Land and Credit Institutions. 2014.

Vol. 72 Beatrice Wambui Muriithi: Commercialization of Smallholder Horticultural Farming in Kenya. Poverty, Gender, and Institutional Arrangements. 2014.

Vol. 73 Christian C. W. Grovermann: Assessment of Pesticide Use Reduction Strategies for Thai Highland Agriculture. Combining Econometrics and Agent-based Modelling. 2015.

Vol. 74 Dawit Diriba Guta: Bio-Based Energy, Rural Livelihoods and Energy Security in Ethiopia. 2015.

Vol. 75 Tigabu Degu Getahun: Industrial Clustering, Firm Performance and Employee Welfare. Evidence from the Shoe and Flower Cluster in Ethiopia. 2016.

Vol. 76 Abu Hayat Md. Saiful Islam: Impact of Technological Innovation on the Poor. Integrated Aquaculture-Agriculture in Bangladesh. 2016.

Vol. 77 Christine Husmann: The Private Sector and the Marginalized Poor. An Assessment of the Potential Role of Business in Reducing Poverty and Marginality in Rural Ethiopia. 2016.

Vol. 78 Qiu Chen: Biomass Energy Economics and Rural Livelihood in Sichuan, China. 2018.

www.peterlang.com

www.ingramcontent.com/pod-product-compliance
Ingram Content Group UK Ltd.
Pitfield, Milton Keynes, MK11 3LW, UK
UKHW021829210426
5322IPUK00004B/95